INTERDISCIPLINARY APPROACHES IN MANAGEMENT EDUCATION

INTERDISCIPLINARY APPROACHES IN MANAGEMENT EDUCATION

Edited by

Tanusree Chakraborty, PhD

Ramandeep Saini, PhD

Rajiv Khosla, PhD

Nandita Mishra, PhD

Ashok Natarajan, PhD

AAP | APPLE ACADEMIC PRESS

First edition published 2025

Apple Academic Press Inc.
1265 Goldenrod Circle, NE,
Palm Bay, FL 32905 USA

760 Laurentian Drive, Unit 19,
Burlington, ON L7N 0A4, CANADA

CRC Press
2385 NW Executive Center Drive,
Suite 320, Boca Raton FL 33431

4 Park Square, Milton Park,
Abingdon, Oxon, OX14 4RN UK

Library and Archives Canada Cataloguing in Publication

CIP data on file with Canada Library and Archives

Library of Congress Cataloging-in-Publication Data

CIP data on file with US Library of Congress

ISBN: 978-1-77491-646-9 (hbk)
ISBN: 978-1-77491-647-6 (pbk)
ISBN: 978-1-00349-549-9 (ebk)

About the Editors

Tanusree Chakraborty, PhD
Faculty, Centre for Management Studies, Administrative Staff College of India (ASCI), Hyderabad, India

Tanusree Chakraborty, PhD, MBA, is a member of the faculty of the Centre for Management Studies at the Administrative Staff College of India (ASCI), Hyderabad, India. With over 20 years of academic and business experience, she has been associated with several business schools and universities in India. Having a thoroughly good academic track record with university ranks, she was formerly a research fellow of the Indian Council of Medical Research. She has around 60 publications to her credit in reputed journals and books, among which are the *Internet Research Journal, the International Journal of Human Resource Management,* and the *International Journal of Productivity and Performance Management.* She has presented over 70 research papers at international conferences, where she has been awarded best paper awards many times. She has conducted management development programs with the state productivity councils and national productivity councils and has trained practicing managers of several public and private sector companies in the areas of behavioral sciences and human resource management. As a psychologist, she has conducted psychometric testing with over 12,000 employees. As a trainer, she has trained over 15,000 employees during her career and has been a resource person for many prominent Indian banks (State Bank of India, Canara Bank, Central Bank of India, Punjab National Bank, Axis Bank) and other companies, including Indian Oil, Exide Industries, Power Grid Corp, Department of Science and Technology of the Government of India, National Institute of Communications Finance, among others. Dr. Chakraborty has published several books by international publishers and has several in the works at present. She has been a panel member in admission and recruitment processes and has conducted GDPIs at academic institutes and corporate. In addition, she has also been a regular faculty for management institutes of repute, teaching MBA and Post Graduate Diploma in Management programs. Dr. Chakraborty holds a

PhD in Applied Psychology from Calcutta University, an MA (Applied Psychology) with specialization in Organizational Psychology, a graduate diploma in Management, and MBA (Human Resources).

Ramandeep Saini, PhD
Director, Chandigarh Business School of Administration;
Dean, International Affairs at Chandigarh Group of Colleges (CGC),
Landran, Mohali, Punjab, India

Ramandeep Saini, PhD, is currently Director of the Chandigarh Business School of Administration and Dean of International Affairs at Chandigarh Group of Colleges (CGC), Landran, Mohali, Punjab, India. She has extensive experience in institution building and specializes in collaborating with international universities and colleges in various domains. She has been awarded a Global Women in Leadership Award 2022 by the Global Council for the Promotion of International Trade. Dr. Saini is an effective speaker and has been invited by international universities to deliver academic lectures and organize workshops on Indian culture. She is recipient of a United Nations Institute for Training and Research (UNITAR) fellowship and is also the founder of Chandigarh Group of Colleges (CGC) International Model United Nations, an annual flagship event of CGC. She is also a Mentor of Change with Atal Innovation Mission, Niti Ayog, Government of India. Dr. Saini has to her credit 17 plus years of experience, which is a mix of corporate and academics. Her areas of expertise are human resource development, organizational behaviour, training and development and organizational development. She has authored one book on human resource development practices and published research papers in national and international journals along with editing and reviewing several books. Dr. Saini has been bestowed with a Best Teacher Award and an Award of Excellence multiple times on Teacher's Day. She has organized various conferences, faculty development programs, and HR summits along with being a resource person in various FDPs.

Rajiv Khosla, PhD

Associate Professor, DAV Institute of Management, Chandigarh, India

Rajiv Khosla, PhD, is currently working as an Associate Professor at DAV Institute of Management, Chandigarh. In addition to holding postgraduate degrees in Economics and Business Administration, he has obtained a PhD in the field of Economics. With 22 years of teaching experience, he has also completed a certification in Public Policy Analysis from the London School of Economics and Political Science (UK). He has authored more than 50 research papers and three books, establishing himself as a prolific writer who regularly contributes to leading Indian business newspapers.

Nandita Mishra, PhD

Director, Chetana's Institute of Management and Research, Mumbai, India

Nandita Mishra, PhD, is currently the Director of Chetana's Institute of Management and Research, Mumbai, India. She has more than 25 years of teaching experience in education and research. She has wide experience teaching managerial economics and economic environment for business, business ethics and corporate governance, corporate grooming, and general management subjects to management students in India and abroad. She has spearheaded several academic administration, industry-institute collaboration and management development programs. She has been the resource person and program director in more than 50 management development programs, and has trained senior-level and middle-level managers and faculty in various programs in behavioural science, managerial effectiveness, CSR, etc. She is a certified trainer in CSR from the Indian Institute of Corporate Affairs. She is an active researcher in the areas of management, with her main research interest areas being leadership, organizational behavior dynamics, consumer behavior, and economic environment of business. She is a nominated member of the Higher Education Council of the Women Indian Chamber of Commerce and Industry (WICCI). She holds a PhD in Management (Jankidevi Bajaj Institute of Management Studies, Mumbai) and an MA and MPhil in Economics and was a gold medalist at both undergraduate and postgraduate levels.

Ashok Natarajan, PhD
Manager, Consultant and Teacher in Accounting, Finance and Corporate Strategy

Ashok Natarajan, PhD, has over 15 years of experience in the academic arena along with more than five years of experience as a practicing manager, both in the USA and India. He has also been a consultant on several business valuation projects. His area of specialization is accounting and finance, and he has published several research papers in top accounting and finance journals. His current research interests include accounting conservatism, business valuation, earnings management, and industrial organization. In addition, he works in the areas of corporate governance and corporate social responsibility in the role of accounting, finance, and corporate strategy. His other interests include curriculum development for management education, leadership in business schools, academic mentoring, as well as the relationship between research output and academic quality. His teaching portfolio includes several courses in accounting, finance, corporate strategy, and allied areas. He has also won numerous awards for his teaching and has been recognized for his service and mentoring roles.

Contents

Contributors

Vikas Arya
Assistant Professor–Marketing, Rabat Business School, International University of Rabat Salé, Morocco

Gaurav Ashesh
School of Business, AURO University, Surat, Gujarat, India

Renu Bala
Assistant Professor, Jan Nayak Ch. Devi Lal Vidyapeeth, Sirsa, Haryana, India

Simerjeet Singh Bawa
Associate Professor, Chandigarh Group of Colleges, Landran, Mohali, Punjab, India

Indu Chauhan
PhD Scholar, Department of Economics, Kurukshetra University, Kurukshetra, Haryana, India

Chetna
PG Department of Mathematics, M.M. Modi College, Patiala, Punjab, India

Erdem Utku Eke
Assistant Professor, Nigde Omer Halisdemir University, Department of FEAS, Nigde, Turkey

Ahmet Arif Eren
Associate Professor, Nigde Omer Halisdemir University, Department of FEAS, Nigde, Turkey

Yusuf Hassan
Assistant Professor–HRM, University of Birmingham, Birmingham, United Kingdom

Navdeep Kaur
Research Scholar, School of Management Studies, Punjabi University, Patiala, Punjab, India

Tanveer Kaur
Research Scholar, University School of Business, Chandigarh University, India

Rajiv Khosla
Associate Professor, DAV Institute of Management, Chandigarh, India

Amit Kumar
RICS School of Built Environment, Amity University, Noida, Uttar Pradesh, India

M. Dileep Kumar
Deputy Vice-Chancellor and Professor, Faculty of Management Sciences,
Nile University of Nigeria, Abuja, Nigeria

R. K. Mishra
Senior Professor, Institute of Public Enterprise, Hyderabad, Telangana, India

Soumita Mitra
Senior Research Scholar, Department of Humanities and Social Science,
IIT (Indian School of Mines), Dhanbad, Jharkhand, India

Pallavi Pahuja
Assistant Professor, Chandigarh Business School of Administration, Punjab, India

Kirti Prashar
Assistant Professor, Sharda University, Greater Noida, UP, India

Geetha Rajaram
Principal and Director, Global Institute of Management Sciences, Bangalore, Karnataka, India

Priya Sachdeva
Associate Professor, Amity University, Noida, Uttar Pradesh, India

Ameet Sao
RICS School of Built Environment, Amity University, Noida, Uttar Pradesh, India

Shulagna Sarkar
DCM/HR, Learning and Development, NLC India Limited, Neyveli, Tamil Nadu, India

Arushi Sethi
Assistant Professor, Chandigarh Group of Colleges, Landran, Punjab, India

Puneet Kumar Sethi
Assistant Professor, Chandigarh Group of Colleges, Landran, Punjab, India

Padmakar Shahare
JAIN (Deemed to be University), Bangalore, Karnataka, India

Anshuman Sharma
Assistant Professor, College of Business Administration, Ajman University, Dubai, UAE

Dhiraj Sharma
Faculty Member, School of Management Studies, Punjabi University, Patiala, Punjab, India

Minakshi Sharma
Research Scholar, Master of Public Health, University of North Florida, Jacksonville, Florida, United States

Ruby Sharma
Assistant Professor, Chandigarh Business School of Administration, Punjab, India

Amrinder Singh
JAIN (Deemed to be University), Bangalore, Karnataka, India

Garima Singh
Student, Institute Chandigarh Business School of Administration, Punjab, India

Geetanjali Singh
Assistant Professor, P.G. Department of Economics, Mehr Chand Mahajan DAV College for Women, Chandigarh, India

Jaskirat Singh
Research Scholar, Punjabi University, Patiala, Punjab, India

Sultan Singh
Professor, MMIM Maharishi Markandeshwar (Deemed to be University), Mullana, Ambala, Haryana, India

Vishal Srivastava
JAIN (Deemed to be University), Bangalore, Karnataka, India

John Evar Strid
Associate Professor, Northern Illinois University, Oak Park, Illinois, United States

Deepti Tara
Freelancer in Management and IT, Ontario, Canada

Preface

The future of work has already arrived. A key point to note is that tasks where management graduates retain their competitive advantage include managing, advising, decision-making, reasoning, communicating, and interacting. The need for management graduates and management education will also be on the rise. One of the trends that has shown results in the past few years is taking an interdisciplinary approach to management education. The New Education Policy 2020 (NEP) aims to implement a multidisciplinary approach across different streams of management education. This has led to innovative management approaches that change the pedagogy of management education.

This book attempts to take an integrated approach to the new paradigm of management education, where both the teacher and the student understand the interdisciplinary approach between ideas and concepts across different domains of study. B-Schools or Management Institutes have been providing professionally trained graduates to meet the requirements of various industries across different sectors. These sectors face complex challenges and uncertainties. Hence, B-Schools should put their best efforts into training students for the complex challenges in business.

The book presents 16 chapters on various themes to consolidate the understanding of interdisciplinary management education. We thank our reviewers for their insights and comments. The editors sincerely invite feedback and suggestions from readers to improve the input quality. We hope that this book enriches its readers with contemporary areas of interdisciplinary management education and worldwide best practices. Wishing readers an enjoyable reading experience!

—*Editors*

Introduction

An interdisciplinary approach in education, although relatively new, has its origins in the idea of a university, continuing with the concept of a liberal education, and finding its echoes in the current notion of a general education. In management education, however, the foundational disciplines were commerce, mathematics, economics, psychology, and sociology. The later inclusion of areas such as organizational behavior, marketing, management strategy or policy, and entrepreneurship indicates the beginnings of an interdisciplinary approach in management education as well. Academic theorists often lament the absence of interdisciplinarity in the field of management education, while others point out that attempts to foster such an approach have often failed. Thus, the idea of an interdisciplinary approach in management education is far from a settled issue.

It can be argued that the development of separate disciplines in management education led to the increasing specialization of knowledge, resulting in the disciplinary structure becoming more rigid, and the reward system for publications reinforcing this insular approach by the creation of new scholarly journals that promote discipline-based research. These are also a natural result of the formation of academic identities that are formed with the institution of management and business schools in universities. For both management theorists and business leaders, this increased focus on specialization in the management education field became a cause for concern as it produced individuals who were skilled in one discipline but were unable to deal with the changing needs of the business they were trained for. The need to counteract this specialized breed of managers led to the formation of interdisciplinary, multidisciplinary, and transdisciplinary approaches in management education.

It is important to distinguish between these approaches. A multidisciplinary approach consists of the coexistence of several disciplines in a program or course. In such an approach, there is hardly any interaction among the disciplines. This has resulted in the introduction of various programs with courses being team-taught by faculty from different disciplines. On the other hand, a transdisciplinary approach aims to demolish disciplinary boundaries and bring in new forms of learning required to solve complex problems, which in turn requires cooperation among faculty from various disciplines. Sometimes, other stakeholders such as practicing managers are brought in to provide insights on the issues and problem-solving approaches. In contrast, an interdisciplinary approach involves two or more disciplines coming together in research or teaching to produce a result that could not have been achieved by any of the disciplines separately. This integration of disciplines has been considered a gold standard for management education, as it involves moving away from a unidimensional approach and adopting a multi-dimensional approach to decision-making and problem-solving.

Over the years, calls for interdisciplinary approaches have led to the emergence of a disciplinary paradox. The paradox is that in order to establish an interdisciplinary field, strong disciplines are needed. Thus, it can be argued that interdisciplinary approaches will take hold and be established when there are existing disciplinary structures in place. However, the existing strength of disciplinary silos represents a major hurdle to an interdisciplinary approach. We have already discussed the establishment of an academic identity, values, and behavior that are created and reinforced by disciplinary institutions. On the contrary, it has been argued that an interdisciplinary approach can strengthen the concept of an identity based on an academic discipline and, in fact, can help create multiple disciplinary identities. Some research has found that interdisciplinary research teams find motivation in the interdisciplinary environment they are in and feel stimulated by the interdisciplinary approach they are utilizing to solve research issues.

Perhaps another argument for an interdisciplinary approach can be found by considering the fields of knowledge. Traditionally, it has been held that there are four fields of knowledge: hard pure, soft pure, hard applied, and soft applied. Academic theorists and leaders have long held that management education, by its very nature, is soft applied knowledge. This outlook has been reinforced by the traditional epistemological approaches used in management education, such as case studies and the trend of reinforcing professional practice. It has also been held that management programs were too vocational and lacked the kind of rigor expected in an academic program. This, in turn, led to a lack of respect for the management school and a lower regard in the eyes of university leaders. More recently, it has been argued that the multidisciplinary nature of management implies that it contains all four types of knowledge fields. This has led to considering the field of management education as a discipline in its own right and has led to calls for an interdisciplinary approach in the field itself.

In the past few years, an interdisciplinary approach to management education has been recommended as a solution to various problems that have arisen in the field. As previously mentioned, the reintroduction of the interdisciplinary approach into the curriculum was driven by the need for management graduates to possess more than just specialized knowledge in one discipline, in order to meet the demands of employers. Employability became a key criterion for evaluating the effectiveness of incorporating an interdisciplinary approach in management schools. Another significant development in recent years has been the increasing use of technology in both campus and industry settings. This has led to a different form of interdisciplinary approach in management education, driven by digitalization and the use of computers in classrooms and industry. Sustainability is another factor that has emerged in the past few years, with the introduction of the Sustainable Development Goals by the United Nations. The management curriculum has been redesigned to incorporate sustainability through an interdisciplinary approach. Other factors that have emerged in recent years include concepts like gamification, the role of instructors, the attitude of students towards interdisciplinary methods, and the impact of the COVID-19 pandemic on reshaping management education.

This book covers all of these themes in its chapters. Employability, digitalization, digital technologies, and sustainability are all addressed in this book, providing an introduction to these factors within an interdisciplinary framework and offering insights into the future of management education, given the growing need for interdisciplinarity. We believe that these chapters will be valuable to readers, whether they are academics or practitioners.

CHAPTER 1

How Does Digitalization Aid in Bringing an Interdisciplinary Approach to Management Education?

TANVEER KAUR,[1] KIRTI PRASHAR,[1] SIMERJEET SINGH,[1] and MINAKSHI SHARMA[2]

[1]*Chandigarh Group of Colleges, Landran, Punjab, India*

[2]*Research Scholar, Master of Public Health, University of North Florida, Jacksonville, Florida, United States*

ABSTRACT

The introduction of information technology has impacted the teaching learning processes and methods. In management education (ME) also the impact of information technological tools can be seen prominently. As ME is directly related with businesses. Any change in business processes and approach will directly affects the ME. This chapter aimed at studying the impact of digitalization on ME and to study how digitalization has helped the management institutions in improving their teaching learning processes. This chapter has highlighted the impact of digitalization on ME, various digital used by different management institutions, challenged faced by institutions in implement digitalization in India, benefits of digitalization of management institutions and future scenario of digitalization of ME in India. This chapter will provide inputs in developing strategies for effective implementation of digitalization in Indian ME sector.

Interdisciplinary Approaches in Management Education.
Tanusree Chakraborty, Ramandeep Saini, Rajiv Khosla, Nandita Mishra, and Ashok Natarajan (Eds.)

1.1 INTRODUCTION

With the advancement of technology, society and the professional world evolve over time and change. Nearly every element of our life has been impacted by the digital revolution. India is the world's second-largest Internet-using country after China. In India, more than 4.66 billion public using internet, and more than 5 billion people are anticipated to own a mobile device. These digital transformation trends have had a significant impact on the education sector as well. Increased competitiveness, globalization, persistent economic uncertainty, shifting demographics, and a rising reliance on technology are all driving changes in management education (ME). Technological advancements and the rise of social media have changed the education dramatically. Rapidity and alertness are the two main features associated with today's digital universe. Adaptive learning has become more widespread as a result of the digital revolution in education. Connectivism, a new learning philosophy, has even emerged (Siemens & Downes, 2008). Students can now self-learn basic ideas and frameworks at any time and from any location thanks to interactive learning tools. Rather than the usual 'chalk and talk' lecturing, with the help of this now faculty can directly interact with their students more effectively through conversations and deliberations. Teaching without technology is now unthinkable, not just because it is an integral part of our daily lives, but also because digital technologies may significantly boost learning when utilized correctly. This chapter has studied the interdisciplinary approach of ME.

1.2 PROBLEM STATEMENT

Technology must support and empower better teaching and learning. Access to world class quality education and providing learning opportunities through digital platform for the management students is the main concern of the management institutions. The study was conducted to study the impact and use of information technology on teaching and learning of education in management.

1.3 METHODOLOGY

The aim of this chapter is to highlight the impact of digitalization on ME and how it is helpful in improving teaching learning process. This is qualitative exploratory research study. The data is gathered from various published secondary data sources like website, research papers, books, newspapers, and articles.

1.4 LITERATURE REVIEW

Christopher Grey (1995) performed the study on ME as a polemic. The study is basically a polemic against the basic structure of ME. The study argues that the essence of management is treated to be in its characteristics and theory, ignoring the critical research and implications. The study depicts the importance of techniques in application of ME. The study further explored the paradoxes and problems in ME and provided insights to the quality initiatives which would help in raising the standards of ME.

French, Robert et al. (1996) in their study compiled the previous studies related to ME and brought together the various approaches of ME, further analyzing their practical implications. The study was conducted in order to bring innovation in ME practices by rethinking the management curriculum and existing practices. The study provided readers with the insights to the critical classroom and latest methodologies which can be the future of ME.

Ben Williamson (2016) conducted the study on digitalization in education and educational governance. The study included the role of management and governance of institutions, educational systems, and individuals through the implementation of digital tools in management and education. The study aimed to analyze the software packages, digital technologies, database platforms and their role in digital education and management. The study concluded that digitalization would play a transforming role in the ME and the governing trends.

Joao Almeida et al. (2019) conducted the study to examine the future of ME. The study aimed to explore the importance of education in entrepreneurship and junior enterprises in ME. The study was conducted on the students from the Portuguese universities enrolled in the management and economics courses. The data was collected from the sample of 139 students. The study revealed that the relation between student enrolled for junior enterprises course and perceived behavior control was found significant. Entrepreneurial intention in these students was greater than the students enrolled for basic management courses. The study revealed the need for integration of extracurricular activities in the basic management curriculum.

1.5 MANAGEMENT EDUCATION (ME)

Students are taught to be corporate leaders, managers, and administrators through ME. ME is considered to be a discipline of higher education (HE). It teaches how skillfully the resources like men, material and machinery can be utilized efficiently and effectively so that organizational and individual goals can be attained successfully. It mainly emphasizes on providing education to handle the responsibilities as executive or administrator of a business organization. It also provides training that how to take work tactfully from people in the organization at any level and how to collaborate different levels of management in an effective manner so that they work in an enthusiastic manner.

Business education is multidisciplinary course. It is that branch of education which imparts knowledge and develops the decision-making capabilities of the learner and also train in dealing with a difficult situation and providing an effective solution to the problem. It also enhances planning skills and then implementing those plans in an effective manner. It also takes inputs from entrepreneurial skill development by educating how to start your own venture and what are the various skills required to start your own business and what are the various sources government and non-government from where funds can be raised to start the new business. It also provides knowledge regarding what are the various schemes of the government which support new startups. In management educational program various disciplines like human resources, marketing, operational management and financial management, information technology, hospitality management and service industry management fields are included to train the learners about all these areas so that they can develop their skills according to their interest and scope of employment opportunities. As after LPG (liberalization, privatization, and globalization) the market has expanded and providing various opportunities in different areas not only in local market but also in global market. Now due to advanced means of communication and transportation now job market is not restricted to national boundaries, but students can exhaust the global opportunities also to use their talents and key skills. Now as per the market trends and arising opportunities in different fields of specialization now the students can choose any specialization of management program.

1.5.1 MANAGEMENT EDUCATION (ME) IN INDIA

With India's exposure to LPG, ME has extended a lot of traction in recent years. After globalization, many MNC's have opened their startups in Indian market thus creating lots of opportunities for business managers. Hence the demand for skilled personnels has increased tremendously. So, there is a need arise of reputed institutes which can impart quality education to fulfill these skilled manpower requirements. Earlier there were a few names which were providing quality education, but now various private institutes are providing quality education in management field few of them are like XLRI Xavier School of management, SPJIMR, FMS-Delhi, NMIMS school of business management, MDI-Gurugram, and many more.

Now the new education policy is also favoring the ME in terms of reach to world class quality. New education policy is emphasizing more on digital tools with the aim of making education more available and affordable. Digitalization of education encourages the leading international universities to establish hubs in India so that Indian students do not need to travel overseas to receive world-class education and experience.

1.6 DIGITALIZATION AS AN AID TO INTERDISCIPLINARY APPROACH TO MANAGEMENT EDUCATION (ME)

An interdisciplinary approach employs two or more disciplines to answer any educational problem, such as educational philosophy, educational psychology, and educational sociology, among others. The interdisciplinary approach brings together the expertise of two or more fields to address a similar challenge (Chakraborty et al., 2021). It enhances research outcomes, increases students' mental efficiency, reduces subjectivity and departmental bias, supports and creates new prospects for additional study, encourages critical thinking, and aids in the connection of ideas.

This shift toward more multi- and interdisciplinary program material affects more than just business institutions. The same trend exists in the opposite way, with non-management institutions increasingly providing their students with the opportunity to learn about business. In its unique and innovative master's program, the Berlin-based Hertie School of Governance, for example, mixes non-business subjects such as data science, public policy, and management studies. CentraleSupélec, a French engineering school, offers a master's degree in data science and business analytics in collaboration with ESSEC Business School.

Academia's movement toward a more multi- and interdisciplinary approach to teaching and learning may be the consequence of increased demand and feasibility as a result of both HE's and society's continuous profound digital transformation, which has been considerably expedited by COVID-19. One may argue that the digitization of business schools (and the corporate world) empowers and entails inter- and multidisciplinarity.

Rapid technology advancement, particularly in artificial intelligence (AI) and robots, causes labor markets and requirements to evolve at the rate of smartphone upgrades. Some jobs will exist in the future that we are unaware of. Employees will be required to (quickly) adapt to new realities and work environments. A multidisciplinary education teaches a wide range of vocabularies and languages from other professions, allowing and supporting people to study autonomously whenever the workplace requires it. When searching on Google or YouTube, knowing the underlying ideas and concepts can assist them discover quality information for re- or upskilling.

Thanks to technology, students have access to a wealth of internet resources, which motivates them to conduct research and therefore become more self-sufficient. It also facilitates learning by making concepts clearer, such as by using an instructional film.

With the introduction of technology in the classroom allows teachers to experiment with different pedagogies and receive immediate response from students and it also make them more active and engaging. Apps and e-textbooks offer a well-organized learning environment. Now days, there is no dearth of digital resources, as they are available to everybody at any time. To make the session more engaging, some educators use gamification. Healthy competition is encouraged through gamification and role-playing. Automation of many tasks like keeping track of the students and their assignments, their attendance, and their performance, saves a lot of time for everyone.

Because of the turbulence and diversity of the environment, it is extremely necessary for institutions participating in ME to implement innovative and radical process-innovations in order to support a structured approach to achieve educational quality.

1.6.1 DIGITALIZATION OF MANAGEMENT EDUCATION (ME)

Management education (ME) improves the skills of the management students to handle the business in a very effective and efficient manner. It helps the owners and business managers in logically analyzing the data gathered about market, customers, and competitors and using this information in taking strategic decisions. It also helps in making effective plans on the basis of predictions made by using this data and implementing them successfully. Now a days the market is data driven. The more the updated data is available with the organization more will be the chance of success of that organization. So, in this digitalization age everything is available online, now it depends on the skills and talent of the managers that by what means they are gathering information from the market. If a manager will not update their self-according to the required market conditions, then that manager cannot be successful. He must be updated towards what are the various advancement has occurred in the information technology and how it can be exploited to be on the top. Now a days every business whether is small size organization or a large size organization are bringing their businesses to digital platform. Digitalization is basically converting material, information, processes and methods into digital forms. Digitalization has created various opportunities as well as posed various threats towards businesses. Now due to digitalization abundance of information is available to the customers before purchasing anything. They make rational buying. Before purchasing any product, they make lots of comparisons like different offers, prices, features, and good will of the brand. So, this has now become very challenging for businesses to project their positive image about their products and services. That is why digitalization has taken a very important place in ME. A management student should be well versed with various information technological tools, their benefits, and their application in different areas of the business. Digitalization helps in gaining competitive advantage by doing things in a cheaper and faster ways than your competitors. Digitalization has opened up new market for the businesses and also improved the production and processes of the business. Digital revolution has changed the business model completely. It has changed the way we communicate and consume various goods and services. Due to digitalization the whole world is now the marketplace for the companies. Now it depends on the skills and resources of the companies that how they can expand their business to global place. Now whatever is the size of the company they should present their product and services in a very convincing way. They need to be more strategic and efficient in their services and presentation. To be successful now businesses need to understand the benefits of digitalization and should move further to exploit them. With the help of digitalization now the business can improve their productivity, positioning, and communication with the different stakeholders.

Because of the uncertainty and unpredictability of the environment, it is extremely desired for the institutions participating in ME to implement innovative and radical system in order to support a systematic approach for improving educational quality. Organizations are required to use evolutionary and revolutionary content learning integration throughout the process. In light of the current situation, which has compelled institutions to embrace digitization, it is essential to regularly rethink the systems and procedures within institutions.

Not only that, but we consider that digitalization should be used as an enabler and facilitator for the development of quality and distribution of education.

1.6.2 INNOVATIVE TOOLS FOR MANAGEMENT EDUCATION (ME)

Management education (ME) in India has still been in its initial stages and need to be revolved completely. The study based on theoretical practices is still taught by various universities using the basic classroom teaching methodology and techniques. The world has changed rapidly and so is the ME field. There has been development of various innovative techniques in the management field which can help in better engagement and learning of students and also providing them with the practical exposure for better understanding. Management institutions use various interactive tools like augmented/virtual reality (AR/VR), gamification, and AI. With this teacher can utilize their time for practical discussions rather than 'chalk and talk' lecturing.

Listed below are various innovative tools used in ME:

1. **Gamification:** It is a process sing which a routine classroom can be turned into the game room. This strategy initiates the practice of imagination and brainstorming among students where they can be engaged into solving the real-life problems. This technique adds the fun element in the teaching and keeps the students active

and involved in the classroom. Students can be taught lectures using various games which will enable them to grasp and execute key fundamentals in functional areas.

2. **Chat Bots:** These are designed with the help of AI as a substitute to the teachers. They can help in clarifying student's doubts, and also conduct automated quizzes and test for students in order to assess their learning but these technologies fail to develop a sense of responsibility and moral values among students.
3. **Video Cases:** Today's generation is not bounded by the time based 9–2 classroom learning. They are more active and are engaged in self-learning outside classrooms with the help of cases studies, recorded video lectures, animated videos, voice enabled chats, discussion-based questions, etc., in order to enhance their overall knowledge and for their overall development and growth.
4. **Simulations:** These are the techniques which are used to design a kind of prototype for the real-life problem and students are asked to analyze the problem, define the issue and the find out the solutions which can be implemented in real world keep in view the real business environment. This technique provides practical exposure to the students. Technology can be embedded with the simulation techniques in order to enhance the performance and interactivity.
5. **Learning Apps:** Many management organizations have also introduced their mobile apps in order to support E Learning and distance learning. The apps are provided to students so that they can easily access the course material, perform assessment test and their academic performance can be easily monitored.
6. **Google Forms and Other Google Tools:** Google tools like Google forms, Google sheet, Google docs had also made the management and collection of data easy. Information can be easily collected from the students and can be compiled with least efforts to provide immediate response for many issues.
7. **Field Education:** Management education (ME) has now not limited itself to the classroom studies. Institutes have also started having tie ups and collaborations with many industries in order to give on field exposure to the management students.

8. **Semester Exchange:** Management education institutes also have tie ups with international affairs in order to provide foreign exposure to the students. Students can opt for the semester exchange program where they can continue one semester of their management graduation from foreign universities.
9. **Smart Card Solutions:** Students not only study about technology and advancements they also apply them in their real situations. Smart card solutions are provided to students where they can order food from canteen or pre book seats using their smart card payments and apps.
10. **E-Books:** Many E-books have also been available now for the management studies at free or very low prices which help in increasing the scope of education by referring more books online instead of one curriculum book.
11. **Patents:** Management students are no more behind in innovations and inventions. They have also started running their entrepreneurial ventures at very young age and also filling patents for their innovative ideas.

    Although all these tools are in the introductory stage of adoption but very soon, they will become the important part of teaching learning system. In the days to come every teacher and student would interact in a virtual mode. And very soon they both will adapt their self into this new virtual mode and will make the best use of these advanced tools of teaching and learning. This will save not only their time but will also reduce their efforts towards managing resources, maintaining class notes in time and in a proper format.

1.6.3 ROLE OF DIGITALIZATION IN FACILITATING AN INTERDISCIPLINARY APPROACH IN MANAGEMENT EDUCATION (ME)

1. **Rapid Increase in the Information Being Shared:** Now with the rise of digital tools information is easily available now students need not to spend hours and hours in searching for the study

content. Even for practical learning now they need not to spend long time in laboratories because now there are various simulator software's are available through which they can get practical knowledge in an artificial environment.

2. **24-Hour Availability of Resources:** As every learner has their own pace and ability of grasping knowledge. With the 24 hours availability of digital resources now the students can learn as per their convenience. And their queries get immediately resolved by the teachers. Now with digitalization anytime anywhere learning is possible. Teaching and learning process is customized according to the learning pace of the students. Online practice tests are available through which students can enhance their skills by doing more and more practice.
3. **New Sources of Revenue for Businesses:** This new e learning methods have opened up new avenues for the startups. Now the businessmen are getting new opportunities to invest in education sector. In digital startups the investment cost is less, and ROI is more.
4. **Education Becomes More Accessible:** With digitalization, now the education from high reputation organization has become accessible. Now the Indian students will also get exposure from the high reputed organizations like Harvard University, Massachusetts institute, university of oxford, Stanford University, etc. Now these institutes are accessible to them without physically moving to foreign land. Because now every organization is looking for international students and are offering online courses.
5. **Reduction of Tedious Tasks:** With digitalization the teachers' tasks has also become easy as now they need not to prepare the content again and again and they need not to repeat the same lecture. Because once recorded session can be repeated as many times as required without wasting time. Now this time can be utilized by the teachers for improving content and generating new resources.
6. Some students who cannot move out of their homes due to family problems now they can continue their studies without physically moving out of their homes. The students from rural areas who

always dream of studying from good colleges in the urban areas now they can enroll themselves in online programs.

7. **Shared Resources:** Now different organizations can do various collaborations by sharing their resources like e-content, teachers, and software's. And good reputed organizations like IIT's and IIMs, etc., can help the growing organizations by attaching their brands with them and can take monetary benefits in return and hence these funds can be used for their infrastructural development.

1.7 CHALLENGES FACED BY DIGITALIZATION IN INDIA

1. **Electricity Problem:** 24-hour power supply is a big challenge for the Indian government in rural as well as urban areas and this is a big obstacle in implementing digitalization policies in education sector.
2. **Internet Connections:** There is still a large portion of the population has no accessibility to internet. The majority residents in urban as well as rural areas still lack the necessary internet capacity, as well as the skills to comprehend digital terms and gadgets.
3. **Internet Connection Plans and the Speeds:** The first factor that springs to our mind when people think about digital education in India is an internet connection. The high-speed internet is still a dream for majority of Indian population. Data packages and their allied costs is a significant barrier for both professors and students, particularly in live classes. Most of the students rely on cellular data which is not sufficient for live classes.
4. **Lack of Devices for Digital Education:** Attending online classes without any digital device is not possible and in India majority of population does not have access to digital devices like computer, laptop, and smart phones. There are many homes which cannot afford even a smart phone to their children.
5. **Shortage of Skill and Experience on Digital Platform:** In India most of teacher are comfortable with chalk and talk learning. They do not feel comfortable while teaching through online mode. And at the same time, they are not interested to teach online.

1.8 FUTURE SCENARIO OF DIGITALIZATION OF MANAGEMENT EDUCATION (ME) IN INDIA

The digital technologies have brought a radical change in the way of delivering education. Now due digitalization quality education is in the reach of everyone. Digital platform has generated opportunities for both teachers and students by offering more participation in the teaching learning process (Chakraborty & Gupta, 2019). The new technology-aided learning tools has changed the way of delivering education in school and colleges both. Online education is the cheaper way of reaching to world-class learning experience. The best thing about digital education is that educator need not to produce the study material again and again. Once created resources can be used many times for future generations. As a result, a significant amount of time and effort is saved. Simultaneously, digital education allows teachers to adapt instructional material to meet the unique needs of individual students.

Digital education is also assisting in the development of the idea of collaborative learning, in which all minds can come together to create a one-of-a-kind learning experience that crosses national borders. With the aim of promoting social inclusion, universities and colleges have introduced digital solutions to promote online education access.

In the near future, interactive digital media will undoubtedly aid in alleviating the problems of education sector in India specially in remote areas the country is facing serious issues, including outmoded teaching methods, a teacher shortage, an inappropriate student–teacher ratio, and a lack of instructional resources.

In India, digital learning will be the most important aspect of future education. It's fascinating to see how smart technologies are altering the nation's national education structure. Digital education is expanding at a fast pace in rural India, and this is because of high-speed internet availability and technological advancements students in villages can learn online and improve their skills and knowledge. Now various startups like study Adda, Byjus, Vedantu, etc., are offering various innovative solutions to educational institutions in India. The Indian government is also changing various educational policies that will help in boosting digital education in Indian market by providing accessibility to innovative educational resources.

It is expected in future that Digital education, like all other disciplines, will witness huge impact on the way universities and colleges give education. We see a plethora of opportunities emerge from digital education to inspire the young people of this nation.

1.9 CONCLUSION

In present time digital tools have impacted almost all spheres of life. It's Influence on education sector is also very prominent. Digitalization has totally changed the teaching perspective. It has come with many advantages like 24 hours learning, removed geographical boundaries, and reduced administrative work and reduced efforts and time saving. But it has also come up with some problems like ignorance of basic human skill, dependency on gadgets, lack of socialization, lack of classroom learning, etc.

As the business world has become very competitive and challenging so for competing effectively and surviving in this turbulence phase revamping of the business education is need of the hour. As ME produces future business managers so to keep them updated and preparing them to deal with this competition there is a need to upgrade their skills as per the business requirements. Managerial expertise with new-age competencies and talents is in high demand in the business world. This makeover is expected to completely restructure the country's business education design, content, and delivery.

The success of any management institution will depend on how fast they accept the changes and adapt their ways according to the changing industry requirements. The institution that will be proactive to the industry needs, will collaborate their resources with other organizations in the form of shared resources and shared infrastructure will become successful. These revolutionary changes will benefit the learner, service provider, organizations, and at last to nation.

To summarize, digital education is the way of the future for India, and it will propel the country to new heights of socioeconomic growth and wealth.

KEYWORDS

- **digital education**
- **digital tools**
- **digitalization**
- **education sector**
- **information technology**
- **management education**
- **student–teacher ratio**

REFERENCES

Almeida, J., Daniel, A. D., & Figueiredo, C., (2019). The future of management education: The role of entrepreneurship education and junior enterprises. The International Journal of Management Education, 100318.

Chakraborty, T., & Gupta, D., (2019). Examining dynamics of emotional intelligence and entrepreneurial intentions among management and non-management students. The International Journal of Indian Psychology, 7(2).

Chakraborty, T., Tripathi, M., & Saha, S., (2021). The dynamics of employee relationships in a digitalized workplace: The role of media richness on workplace culture. In: Critical Issues on Changing Dynamics in Employee Relations and Workforce Diversity (pp. 175–205). IGI Global.

French, R., & Grey, C., (1996). Rethinking Management Education. Sage Publications, 2455 Teller Road, Thousand Oaks, CA 91320-2218 (hardcover: ISBN-0-8039-7782-4; paperback: ISBN-0-8039-7783-2, stock No. 4038; $26.95.

Grey, C., & Mitev, N., (1995). Management education: A polemic. Management Learning, 26(1), 73–90.

Siemens, G., & Downes, S., (2008). Teoría Del Conectivismo. Consultado en: http://www.elearnspace.org (accessed on 20 February 2024).

Williamson, B., (2016). Digital education governance: An introduction. European Educational Research Journal, 15(1), 3–13.

CHAPTER 2

Where Does Digital Technology Meet Management Education? The Present Interdisciplinary Nature of Management Education

RAJIV KHOSLA,[1] NAVDEEP KAUR,[2] and DEEPTI TARA[3]

[1]*Associate Professor, DAV Institute of Management, Chandigarh, India*

[2]*Research Scholar, School of Management Studies, Punjabi University, Patiala, Punjab, India*

[3]*Freelancer in Management and IT, Ontario, Canada*

ABSTRACT

The main purpose of this research chapter is to gain an understanding of the importance and role of digital technology over different sectors worldwide and more specifically in the area of education where it is playing a prominent role in enhancing the skills and knowledge of the individuals involved therein. In the present scenario of pandemic digital technology has been a great help in maintaining the constant pace of the livelihood of most of the individuals engaged in different sectors. As per the estimates of the researchers, the education sector was the main area that was to be affected by the outbreak of the COVID-19 pandemic, but it is because of the digital technology that the education sector remains protected from the outbreak of COVID-19. Digital technology refers to the technology which helps the individual to execute the physical task on a virtual platform. It has its applicability in almost all the work and sector worldwide like

Interdisciplinary Approaches in Management Education.
Tanusree Chakraborty, Ramandeep Saini, Rajiv Khosla, Nandita Mishra, and Ashok Natarajan (Eds.)

School, hospital, workplace, and many more. The most prominent benefit of digital technology is its worldwide reach and the threat accompanying digital technology is the cybercrimes that are in force today. Thus, it becomes our responsibility to use digital technology in a safe responsible and efficient manner so that the best results can be derived.

2.1 INTRODUCTION

In the present era of technology, technology is used by every individual extensively in its day-to-day life. Due to its wide application, it is being used by individuals of all factors like students, working professionals, households, etc. Digital technology opens the world to an individual in a small handset or a system that can provide him all the information in real-time about anything anywhere anytime. Nowadays there are various forms in which digital technology is being used by individuals which mainly include smartphones, laptops, personal computers, and tablets. The day-to-day innovation in the field of Technology has led to a multiform impact in almost all areas. The telecommunication industry is growing rapidly with the innovation in digital technology introducing new and new features making the utility of smartphones more comprehensive. As Castells (2006) acknowledge that technology is not in society it is society, reflects the influence of technology in our daily life.

As Digital Technology has a wide range of impacts and the education sector is no exception, thus, the impact of digital technology can also be analyzed from point of view of the education sector. It can be said that education is one of the greatest sectors impacted by the introduction of digital technology. With the help of various means available to the students in the education sector, their learning is not confined to the schools or University rather now they can learn various other things also as per their interest from the internet anytime as per their convenience. This has been a great value addition to the knowledge of the students where they can acquire a wide range of information just by sitting at home, doing nothing. There are various resources available with respect to the education industry one of which is the gaming software. The resources are quite helpful in maximizing the skills and abilities of the students. As per Mustapha, Van, Shahverdi, Qureshi, & Khan (2021), the students are more likely to develop good analytical and thinking skills with the help of gaming tech in education. The technology has been found to enhance the skills by 88% on an average. However, with these benefits, various other threats are also accompanying digital technology. The three men of these threats include cyber safety, cybercrime, and cyber security. This usually refers to the unauthorized access to the information of an individual without his consent and using the information in an inappropriate manner which is not permissible by law. Due to these threads are comparing the digital technology.

2.1.1 HIGHLIGHTS OF THE CHAPTER

- Education being a primary and most important area which is impacted after the outbreak of technology.
- Traditional aspect of education is now becoming outdated and inefficient.
- Various positive impacts and negative impacts can be identified on technology on the education sector and students.
- Technologies like RDBMS, EDMS are driving the educational sector.
- Student's efficiency has been enhanced and also more career opportunities are developing.
- Threats are also accompanying the education sector with respect to the technological enhancement like, cyber security is a major concern.
- Efficient practices should be followed so that, positive results can be drawn, and success can be derived from technology.

2.2 TRADITIONAL ASPECT OF EDUCATION

Considering the educational sector 30 to 40 years from now in the past there was no search service available which are accessible to the students nowadays. Earlier most of the work was done physically due to lack of technological facilities being available to them. It is mostly characterized with face-to-face interaction between the teachers and students involved. No other means of communication was available due to which the education was limited to the school and college itself. People who want to acquire skills education need to enroll themselves into an institution that offers formal education. Also, the teacher needs to be very skillful to train other students under him and engage the attention of the students by interestingly presenting the content.

With time people started moving from the villages to the industrial or urban areas and therefore the number of rains started to increase in these institutions as a result of which the need for new schools was felt and the expansion of the existing institutions was demanded to cope with the increased number of students. Due to limitations in terms of capital and time, it was not possible to build such a higher number of schools; thus, the need for Digital Technology was felt.

2.2.1 CHARACTERIZATION OF THE EARLIER TRADITIONAL ASPECT OF EDUCATION

1. **Lack of Means to Communicate:** There were limited means of communication was available to the students. The only way with which students can communicate with their respective teachers was the face-to-face interaction which can be done in the limited period that is in working hours and to gain anything extra or to clear the doubts of the students there are required to wait for the next working day.
2. **Lack of Access to Information:** The students do not have any access to the information which is outside the scope of the student's syllabus in which he or she had an interest. In the limited period, only the classroom syllabus was able to be completed and no other interesting and useful information can be communicated to the students.
3. **Access to Education:** The education was limited to only a few children as most of the schools were located in urban areas which were not accessible by the children of rural areas. This problem has been now resolved with the help of technology as now it is education can be accessed by anyone anywhere by just having some basic utilities and material with them with which they can connect with the outer world.

2.2.2 BENEFITS OF TRADITIONAL EDUCATION

Besides several disadvantages of traditional education, there were certain advantages also which is not present in the digital platform. Some of the advantages on basis of which traditional education can be preferred are as follows:

- Face to face communication leading to better understanding;
- No glitch chance as any technology is involved thus, lesser issues in communication;
- Less costly as no expensive equipment is required to be set up;
- More security as the information of individuals can't be stolen and misused by cybercriminals.

2.3 DIGITAL FRAMEWORK WITH RESPECT TO EDUCATION

Digital technology has a wide range of applications in the educational sector. It is through various means that digital technology is playing a crucial role in the segment of education. It targets most of the areas which are required for the efficient performance of the educational sector and on basis of which the success of the sector depends (Chakraborty & Gupta, 2019). There are various benefits of digital technology to education and some of the digital framework areas concerning education as stated below:

1. **Collaboration:** Digital technology helps in enhancing the collaboration between students and the teachers and among students. Collaboration is an important area of cooperative functioning in any organization. Effective collaboration can ensure that all individuals are expanding their efforts toward the achievement of a common goal which helps in promoting teamwork.

2. **Communication:** Digital Technology also facilitates communication. With the help of technology students can interact with various personal and literates anytime and to anyone from anywhere in the world. The problem is the communication process that arises due to different locations of the parties is solved with the help of digital technology. Not only verbal communication but now with about digital technology one can communicate in person also with the use of video calling with several people in a conference.

3. **Enhance Capabilities:** Capabilities of the students is enhanced with the help of digital technology. With the help of information accessible from various sources a student can now increase his knowledge regarding anything he is interested toward; thus, digital technology also plays a great role in enhancing capabilities. With access to different information, the thinking capacity of the individual and hence and more creative ideas can be now developed within the students themselves. As per Underwood (2009), the grades have improved after the introduction of broadband.
4. **Connection:** Digital technology also helps an individual to develop connections which might help them in their career opportunity. Various platforms are now available with which the students can come in contact with the professionals in their respective fields and can take guidance from them which will help them to focus on the things which require many considerations. This is not only the professional network of the world is enhanced but also the career opportunities are increased widely. As per Selwyn (2001), digital technology can be used as a way to reduce the exclusions of some social groups from access to education.
5. **Systematic Approach:** Digital technology promotes a systematic approach towards education. Now the students are no longer dependent on the physical classes to enhance their knowledge, but they can them self-organized a digital platform service with which they can get the required information. Various programs are available on the internet which provides professional qualification or degree courses to the individuals which students should opt for.
6. **Employment Generation:** The inclusion of digital technology in the educational sector has increased the process of employment generation. It does not initiate or benefit the students by increasing their employability by enhancing their skills and knowledge but also has emerged as a separate field for employment. Various software developers are now engaged in the process of developing more user-friendly and easily accessible platforms which can be used for education.
7. **Language Barrier:** With the introduction of digital technology one of the major problems of the language barrier is eliminated.

Now the students can get a personalized experience of learning in a format that is suitable and is understandable to them. Earlier the students have fewer options thus, they were bound to learn from only the teacher which is available in their school but now they can learn on the web in any language which is suitable to them without any additional cost.

2.4 THREATS

1. **Cyber Safety:** When the enhancement of the technology and its wider acceptance by most individuals the cyber safety has become one of the concerns or one of the major concerns of the parties involved. To ensure cyber security various platforms software have been developed which provides the insurance that the safety of the user of Technology will be maintained and will be prevented from any unauthorized access. However, even after these, certain loopholes remain open for the cybercriminals who take advantage of this.
2. **Cyber Crime:** It is an online crime or computer crime that is conducted by use of a computer as an instrument to execute some illegal activities such as stealing identities for violating the privacy of individuals by accessing the information in an illegal way switch is not permissible by law. The most common type of cybercrime includes email and internet fraud.
3. **Cyber Security:** It refers to the various measures undertaken to protect systems networks and programs from digital attacks. Under these various actions are taken to protect the network. To do this various software have been developed which are installed to secure the network. Firewalls are being developed to prevent unauthorized access to the system.

2.5 TECHNOLOGY SUPPORTING THE DIGITAL TECHNOLOGY IN EDUCATION

Various technologies are operating in the area of education. The technological integration is intended to initiate the innovative nature and is likely to prepare the students for the future works and society (Ng, 2016). Frequent modifications are making the technology is more efficient as a result of which there are several technologies in which the selections are to be made. As per Higgins et al. (2012), there are so many technologies that the focus has now been shifted from like word technology to be used to that what type of Technology will be suitable for with educational purpose.

2.5.1 RDBMS

RDBMS stands for relational database management system that is used by its applications to manage operational data. It is a database system that helps in storing information about a particular individual or group of individuals in a tabular format which can be accessed by the person who has the authority to do so. In the relational database management system, the data is stored in a tabular format arranged in rows and columns. It is an efficient system that helps in storing a large volume of data that can be extracted easily as and when required.

It is contributing to the educational sector in the way of storing the data of students which helps the authority to store the student's data and help them to track their involvement and providing all the necessary service which is important for their studies purpose. Whether data the authority can know that which student requires what, so that, it can be provided to him like books and other facilities. In the case of education, the relational database management system is known as the student database management system that is SDMS. It helps to keep a record of the student's fees submission any scheme which is to be provided and the name of students with pending services to be provided to them later and many more.

The biggest advantage of a student database management system is that it is secured through various means like various software are installed in the system to protect it from unauthorized access. It is very hard for hackers to enter into the system without permission and therefore it is a safer way to access the system and keep a proper record of the students.

It helps in better communication. The system enables the teachers to remain in contact with the students and stay connected. Instead of depending only on the manual records, it provides the teachers with the facility to access records anywhere and sends an instant update to the parents as and when required.

2.5.2 EDMS

EDMS stands for an electronic database management system which is mostly similar to the RDBMS relational database management system. It also facilitates keeping a record of the students in the educational sector. The only difference between the relational database management system and the electronic database management system is that RDBMS records data in an online format which can be accessed by any individual having a thought to do so anywhere in the world whereas an electronic database management system is like a manual recording of data with the help of electronic software like a computer, MS Excel, etc.

It has the same advantages as a relational database management system, but the disadvantage is that it can't be accessed from anywhere in the world and it is mostly like manual recording thus, teachers are required to be dependent on the records which are stored at a particular place. Thus, it can be said that there is a lack of mobility in the system but yet it is less expensive and therefore suitable for most of the organizations that are not in the position to do much expenditure like small scale institutions. It is also more secure because of its left presence in the online network it becomes somewhat difficult for hackers to enter into the systems. Thus, it can be given preference over a relational database management system.

2.6 MEANS OF COMMUNICATION IN DIGITAL TECHNOLOGY PLATFORM

With the introduction of digital technology platforms, there has been a change in the mode through which the communication process is executed between the students and teachers and how the lectures and the knowledge are exchanged from teacher to student. The pace of technological change is quite rapid and educational sector is likely to be affected in the same direction positively (Ng, 2016). Earlier there were only theoretical books which were provided to students and were explained by the teachers but with the introduction of digital technology platform, there have been much greater aspects now available with which the lectures can be provided and more efficiently and interestingly.

These ways of providing education include the following:

1. **Graphic:** The students are provided with the facility of understanding the whole aspect with the help of graphics like charts diagrams and other visuals which help the student to better understand the information. The earlier way of reading the text and getting the information was not much efficient as do students interest can't be attracted toward the content of the study. But now as the information is provided with the help of graphics the students are more inclined what understanding the content.
2. **Audio–Visual Means:** These are also a new way with which information is provided to the students. Audio-visual means include the videos animation which is integrated with the online software and are accessed by the students. It is also an interesting way of providing information as students learn faster through this. It often includes videos regarding a particular topic which is presented from point of view of a practical scenario like for the concept of the stock market, the teacher provides the student with the video in which the stock market is explained from point of view of an investor leading to a better understanding of students.
3. **Text:** No doubt there has been an increase in other ways with which information can be provided to the students but yet the text is an inseparable part of the studies. It is important to include text as it is a major part of the content that is delivered to the students. Graphics and other ways can attract the students to word studies but cannot provide as much information as it can be provided by text therefore even in digital studies text is not discarded.
4. **Uses of Digital Technology in Education:** The technology is used widely in the education sector for various purposes. But the use can be broadly categorized as Virtual Classes and Assessment. Both of these are the greatest components of the Education sector. These uses can be further explained as follows.
5. **Virtual Classes:** One of the major uses of digital technology in the field of education is virtual classes. Virtual classes refer to classes that are conducted using an electronic platform. Virtual classes are conducted using digital technology to provide the student with a similar experience as they were having during the normal period or face-to-face interaction between the teacher and students. In

the virtual classes, students are provided with almost the same option as they were provided in the physical classes like having the ability to with the help of mike or presenting something by activating video mode on.

6. **Assessment:** Another thing for which the technology is used in the field of education is for the assessment of the students for which they have been taught. It forms an important part of digital education as an assessment the process of learning is incomplete. There are various platforms available that help in conducting exams in a proctored mode where students are governed by the teachers who are assigned the responsibility to do so that the exams can be conducted efficiently.

2.7 MODES TO ACCESS THE DIGITAL EDUCATION

There can be various ways to access education through digital platforms (Chakraborty et al., 2021). In the era of Technology where new and new platforms are being developed at infrequent intervals, it is quite simple nowadays to access education using digital technology. Platforms like Byju's, Vedantu which are providing institutions or physical like education with the certification also to the students who enroll into their programs and join classes through online means.

2.7.1 WEB/INTERNET

The most common way of accessing digital education is through the web. There are various websites available on which a student can go or surf to get the required information. It consists of a mix of several websites which offer frees education as well as certain website also which are paid. The web can be classified into two types based on the accessibility. It includes the following:

1. **Public Website:** These are the websites that are accessible by all the persons that have access to the internet. It does not require any login ID or password to access the content of the website and can be accessed by any random general person. Most of the sites which are public in nature are free and does not involve any transactions which need to be executed to access the content. It is also known as the internet.

2. **Private Website:** These are the websites which are owned by a specific institution and only the individuals who are part of that institution are allowed to access the content of the website. It requires the usage of login ID and password so that it can be ensured that the person who is accessing the content has the authority to do so. The sites which are private in nature involve some transactions to be executed to enter into the website. It is an example of an intranet that is developed by some institutions specifically for their members.

2.7.2 SEARCH ENGINES

It is also a great tool to access digital education. There are millions of search engines available like Google Yahoo, etc., which can be used to retrieve information as per the requirement of any individual. Based on the query there are various alternatives available to any individual from which the most suitable one can be selected which satisfy the need of the user. It has helped greatly the researcher's teachers and students to acquire information during the pandemic.

2.8 MODES OF SHARING INFORMATION WITH RESPECT TO EDUCATION

Communication is an important aspect of education. It is through communication that the information is shared between teachers and students. This exchange of information between students and teachers can be done through various means which are facilitated by digital technology. With the advancement of technology, the communication process is now in hand and is not limited to place and time. Now anyone can communicate with anyone from anywhere in the world with the help of digital technology. The various means available today include video conferencing, virtual meetings, and chats messages with the help of messaging apps like What Sapp Facebook and many more. These include the following as mentioned:

1. **Text Messages:** There are various apps available today with which text messages can be sent to anyone and they have become a great source of communication is done between two or more individuals. As we're concerned with education and most of the

relations included are taken into consideration are formal in nature thus, the most frequently used text messaging app is Google mail or Gmail. It has become one of the most widely used apps for formal communications and with this single message can be sent to multiple people at the same time with just a click of a button.

2. **Video Conferencing:** Text messages only allow us to send information to the individual but do not guarantee the reply from the other person and also that whether he has properly understood the message or not. This can create a lot of problems in the accomplishment of any task. To avoid this there is a solution available for the communication process known as video conferencing individual can contact each other on a video call and can talk as they are talking as person to person. With this most of the problems can be solved as on the spot query resolution can be done and it can also be ensured that the receiver has properly understood the message and is now well equipped to do a particular work which is assigned to him.
3. **Virtual Meetings:** It has evolved as a new way for professionals to communicate with each other. Now they can conduct meeting with one another through virtual meetings. It can be done between two individuals and more than two individuals as it was done physically in an office room. It is now widely used in the education sector to conduct classes of the students in an online manner. Most of the institutions developed their path by following which the joint their virtual classes sitting at their home whereas some users already established platforms like zoom app to conduct their classes.

2.9 HURDLES IN WAY OF DIGITAL TECHNOLOGY IN EDUCATION

1. **Lack of Resources:** There are not many resources available that can be used to access online education. A lot of money is required to be invested to create a single platform on which education can be accessed by several individuals. Expenditure on Web services satellites and platform creation is required to be done to create an efficient and user-friendly platform that can be used by the individuals as per their needs. However, doing such expenditure

is not possible for all except certain large investors due to which the reach of digital technology in the field of education is very little.

2. **Mentality of Individual:** The traditional aspect of education has an inseparable impact on the mind-set of individuals. More than 92–95% of individuals believe that there is no benefit of the online certifications which are gain by them and therefore very few of them are interested to join online education. It will require a great effort and proper strategy so that the individuals can be persuaded and attracted towards online education so that they can adopt them without any hesitation. As per Ng (2016), even after 5 decades there is still confusion regarding the usage and reluctance to move beyond the traditional aspect.

3. **Student's Attitude:** Most of the students do not take online education seriously and just consider this as a way to time pass. In online education the students by joining the class India and other activities and do not pay much attention to what the teacher is saying. Due to this attitude of students, most of the parents are also reluctant to put their children in online classes knowing that it will not be of any use because of the student's attitude.

4. **Time Lag:** Apart from a huge investment to develop an online platform for an educational facility also take a lot of time to develop such a platform. A lot of research work is required to be e undertaken to find that what system will be more accurate and suitable for different types of customers for individuals who are the target segment of the creators. Due to this time lag in the development of online platforms the progress of digital technology in the area of education is very slow for the past few years.

2.10 RESPONSIBILITIES OF STUDENTS AND TEACHERS

2.10.1 STUDENT'S RESPONSIBILITIES

1. **Create Awareness:** It becomes the responsibility of students to create awareness regarding online education. As they belong to current generation, they know better about the benefits of this

and therefore they should inform other also so that more and more individuals come and join online classes helping them to better develop their understanding and enjoy learning with more facilities.

2. **Change Their Attitude:** Wouldn't they be also required to change their attitudes toward online education? They should consider this as a vast source of information and therefore should take it seriously to gain knowledge rather than enjoying it unnecessarily. The students should consider this as an opportunity to increase their understanding of the concepts as a lot of time and effort is being invested in this.

2.10.2 TEACHER'S RESPONSIBILITIES

1. **Guide Properly:** It is the responsibility of the teachers to guide their respective students correctly toward the usage of the technology in a rational manner that will benefit them and not using the technology for the wrong purpose. This will help the students to better understand how they should use the technology. It will add to the utility of the technology and will benefit the students in a better manner.
2. **Update Them to Adopt Technology:** The teachers should adapt the technology at their part so that they can better use them and can use the technology efficiently so that interaction can be done between teacher and student without any hindrance. For better implementation of digital technology and getting its advantage to the greater extent, the teacher should have their hands on the technology. For this, their institution can also arrange some training programs in which the teacher will get sufficient exposure to the new technology.

2.11 DIGITAL TECHNOLOGY IMPACT ON EMPLOYABILITY

The recent outbreak of COVID-19 has taught us that having skills and knowledge is not sufficient to get the desired work. It has revealed that how technology has influenced the workgroup. Most of the jobs which are available now are technology-based and the persons which are in this sector have more scope of getting employment than any other group of people. During the coved pandemic, most of the working professionals were led out from their jobs by giving reasons of loss that occurred during the Lockdown Period.

As per Frankiewicz & Chamorro-Premuzic (2020), there is no other way to overcome this issue apart from learning how we can co-exist within the digital world. We should drive the change and should consider people first then technology and to make progress slowly up to the greatest what we can do.

2.11.1 PACKAGES BEFORE AND AFTER THE DIGITAL TECHNOLOGY

Usually, there is a negative impact of digital technology. But in certain areas, there has been a positive impact. As per the data by Date to Compete (2021), the last year due to advancement in technology the packages of IIMs have increased by 2 to 3% in 2021. During the previous year despite having a strained economy, there has been a rise of 20% in pre-placement offers. This can be attributed to the need fora person who can efficiently manage the organization during the COVID pandemic. The pay of the IIT has also not been infected by the outbreak of the COVID-19 pandemic (Date to Compete, 2021). As they are mainly concerned with the use of technology, they are more likely to get a job thus, their compensation has not been impacted.

2.12 KEY TAKEAWAYS FROM CHAPTER

- Technology is a great asset if used appropriately. The same is the case with technology used in education. Even in 21st century we have not been able to utilize the technology to its greatest extent. The need for change in practices is required so that, technology can be better utilized in education sector and more appropriate results can be derived.
- Traditional schools that have old tools and techniques like, white-board or no tech, have far less level of productivity and efficiency

than those having well-resourced Technology infrastructure (Underwood, 2009).

- Various positive outcomes are there like, enhancement in skills and abilities, mobile learning, etc. Therefore, technology should be utilized by the various educational institutions so that mass benefits can be expected. The new and more innovative ways should be implemented like, simulations and business games so that the managerial skills can be developed, and students can be made industry ready.
- Various threats are also accompanying the education sector like, risk of cyber theft or hacking. The same should be accounted for and proper measures should be taken so that the risk can be avoided to a significant extent. The problems if solved can attract more students and institutions to adopt the new normal of online education.
- More innovative ways should be identified, and precautions should be taken so that the technology can be implemented in better manner in the education sector. The technology is growing at a fast pace and therefore, more and more tech and ideas are coming day by day which can help to make the system more effective and efficient.

2.13 CONCLUSION

Concluding this, we can say that digital technology has a great impact on every sector and education is no exception. There is a great impact of digital technology on the educational sector, and it has almost revolutionized the educational sector by enabling students from every corner of the world to access any kind of education in which they are interested. The technology is very much efficient in rendering benefits if it is utilized efficiently and is properly used by the teachers and students. Better communication and less cost are the major advantages of digital technology.

However, there are certain problems also associated which needs to be catered to so that more and more students can be attracted towards online education. The teachers and the students both have their responsibility is to be fulfilled so that online education can become a success. As new and innovations are done regular basis therefore there must be some training program in theater both for students and teachers with which they can understand that how that can be used more efficiently. If this is accomplished, then there would surely be the benefit of the technology in the education sector also.

KEYWORDS

- **benefits**
- **digital education**
- **digital technology**
- **education**
- **employability**
- **search engines**
- **threats**
- **traditional approach**

REFERENCES

Castells, M., & Cardoso, G., (2005). *The Network Society: From Knowledge to Policy* (pp. 3–21). Washington, DC: Centre for Transatlantic Relations, Paul H. Nitze School of Advanced International Studies, Johns Hopkins University.

Chakraborty, T., & Gupta, D., (2019). Examining dynamics of emotional intelligence and entrepreneurial intentions among management and non-management students. *The International Journal of Indian Psychology, 7*(2).

Dare to Compete, DTC, (2021). *IIMs and IITs See an Upsurge in Placement Packages During COVID-19*. DARE TO COMPETE. Available at: https://dare2compete.com/blog/iims-and-iits-see-an-upsurge-in-placement-packages-during-covid-19 (accessed on 20 February 2024).

Frankiewicz, B. F., & Chamorro-Premuzic, T. C., (2020). *Digital Transformation Is About Talent, Not Technology*. Harvard Business Review. Available at: https://hbr.org/2020/05/digital-transformation-is-about-talent-not-technology (accessed on 20 February 2024).

Higgins, S., Xiao, Z., & Katsipataki, M., (2012). *The Impact of Digital Technology on Learning: A Summary for the Education Endowment Foundation.* Full Report. Education Endowment Foundation. Available at: https://eric.ed.gov/?id=ED612174 (accessed on 20 February 2024).

Mustapha, I., Van, N. T., Shahverdi, M., Qureshi, M. I., & Khan, N., (2021). Effectiveness of digital technology in education during COVID-19 pandemic. A bibliometric analysis. *International Journal of Interactive Mobile Technologies, 15*(8).

Ng, W., (2016). *New Digital Technology in Education.* Springer International PU. Available at: https://link.springer.com/book/10.1007/978-3-319-05822-1 (accessed on 20 February 2024).

Selwyn, N., (2012). *Education in a Digital World: Global Perspectives on Technology and Education.* Routledge. https://www.routledge.com/Education-in-a-Digital-World-Global-Perspectives-on-Technology-and-Education/Selwyn/p/book/9780415808453 (accessed on 20 February 2024).

Underwood, J. D., (2009). *The Impact of Digital Technology: A Review of the Evidence of the Impact of Digital Technologies on Formal Education.* Digital Education Resource Archive. Available at: http://dera.ioe.ac.uk/id/eprint/10491 (accessed on 20 February 2024).

CHAPTER 3

Can a Holistic Management Education Be Linked to Employability Among Management Students? Teacher's View

RENU BALA[1] and ERDEM UTKU EKE[2]

[1]*Assistant Professor, Jan Nayak Ch. Devi Lal Vidyapeeth, Sirsa, Haryana, India*

[2]*Assistant Professor, Nigde Omer Halisdemir University, Department of FEAS, Nigde, Turkey*

ABSTRACT

The issues related to the employability of students have become a serious matter for society, as after the completion of management degrees, students remain unemployable. This chapter aims to find out what factors are responsible for low employability of management graduates. Data have been collected from 160 teachers at various universities such as central, state, and private. IIMs have also been covered under the sample. After the collection of data, t-test, and ANOVA have been applied to make the comparison between the views of teachers. The results of study revealed that there exist several factors which are highly responsible for low employability of management students. The society cannot blame the teachers only, for the low employability as students and management of higher educational institutions both are equally responsible for such issues. To remove such kind of problems, students, teachers, and the higher authorities of educational institutions should make the efforts in combined way.

Interdisciplinary Approaches in Management Education.
Tanusree Chakraborty, Ramandeep Saini, Rajiv Khosla, Nandita Mishra, and Ashok Natarajan (Eds.)

3.1 INTRODUCTION

At present, the matter of graduate employability has become a key issue for the higher education (HE) institutions (Saunders & Zuzel, 2010) as employability is becoming one of the major outcomes of, HE instead it being weakened conceptualized for the sector (Behle, 2020). But instead, a lot of technological advancements, management graduates are not getting the job in the marketplace. A lot of factors are responsible for the issue of unemployability of management graduates. As employers report that graduates do not take their job serious as they are not prepared for world of work and have lack of basic skills which are essential for the employability (Tymon, 2013). Some graduates think that foreign degree provide the employability to the graduates but in actual graduates with foreign degrees or diplomas have to face various hurdles to enter in the labor market, but after the selection they get better economic rewards as compared to other graduates (Jenssen, 2011). The question arise that what initiatives are taken by the government and higher educational institutions to solve unemployability issues and what they should do for the better future opportunities of students (Frankham, 2017). To solve the issues related to unemployability of management graduates, HE Institutions should work in a joint way to create the awareness in the students regarding required skills for their better future (Succi & Canovi, 2020). As well as educational institutions should concentrate on integration of employability and academic learning by avoiding the focus on theoretical knowledge only (Speight et al., 2013). On other side, students should try to highly engage in extra-curricular activities, so that they can improve their employability skills (Sin et al., 2016) such as teamwork, communication, presentation, and creativity, etc. (Jackson, 2013).

3.2 REVIEW OF LITERATURE

Curtis & Shani (2002) indicated that most of the students join part time jobs with their studies, which effect their academic results badly as students' scoreless marks due to the short lectures and classes, etc., and they learn less due to their timing spend on job, instead of study. Yorke & Thomas (2003) revealed that the students who belong to lower socio-economic groups possess various skills such as sincerity, punctuality, hardworking, and loyalty, etc. But the employers have to offer them various commitments for their retention in the organization as to retain good talent for the success of the company. Pawan et al. (2005) stated that women get less career opportunity in the India due to the socio-cultural, economic, and historical factors as most of the Indian people are stereotypes on gender-basis and they do not allow to women to do job. Jill (2009) stated that the students who perform their duties in the hospitals after the completion of their hospital management courses are not satisfy with their jobs as various unethical, unfair and illegal practices influence them and they do not feel comfortable at their workplace. Pan & Lee (2011) presented that students have a view that employability is associated with employment processes only, which is not right as it includes other various aspects also such as academic publications of the students that provide benefits to the universities and students both. Richardson & Butler (2012) presented that Malaysian undergraduate students do not give importance to the tourism and hospitality stream of management as they believe that this field do not provide good future opportunities to them. Zahid et al. (2013) showed that them exist less difference in the views of managers and students regarding MBA courses. As both accept that less practical knowledge is provided to the students during MBA study. Paadi (2014) identified various skills which are highly demanded by the employers. Employers give more importance to soft skills, but it does not mean that they do not want hard skills in the students. As employers stated that, both soft and hard skills should be possessed by the students to get good job. Aida et al. (2015) examined the worth of employability programs for the students and stated that such kinds of programs effect the perception of students regarding various aspects of employability and suggested that these programs should be organized by the educational institutions on regular intervals. Cheong et al. (2016) revealed that Malaysian students demand less hiring costs as compared to foreign graduates, but they possess less employability skills also. This is the reasons most of Malaysian students are hired on low salary as compared to other country's students. Leach (2017) highlighted the students' perception regarding those factors which effect their career and not under their control. Students think that various economic, social and political policy effect their career opportunities and their outcomes may be positive or may be negative also. Moira et al. (2018) concluded that to achieve good grades in the academic is not sufficient for the better job opportunities as various other factors such as personal skills, self-directed learning and motivation, etc., also play a significant role in the students' career life. William et al. (2019) founded that career ownership, human capital and careers advice are major components of the self-perceived employment and suggested that teachers should try to generate self-perception of employability in the graduates for their better career. Bala & Singh (2020) presented that what skills should be developed in the students by the faculty members so that they can get job in the banking field. The bank officials described that teachers should concentrate on modern employability skills such as ICT skills and social media skills, etc., which are highly required in the banking industry. Bala & Singh (2020) indicated the students' views regarding employability skills. Students presented that faculty members pay focus on good academic score only by avoiding the importance of practical knowledge and experience. Bala & Singh (2021) highlighted the teachers' perception on employability skills of management students. The results of the study revealed that teachers at the private universities pay focus on the modern employability skills and try to generate such skills in the students while teachers at state universities focus on basic employability skills and make less efforts to generate modern skills in the students.

3.3 RESEARCH OBJECTIVES

- To analyze the gender-wise teacher's viewpoint towards factors responsible for low employability of management graduates.
- To analyze the institution-wise teacher's viewpoint towards factors responsible for low employability of management graduates.

3.4 HYPOTHESIS

- **H_{01}:** There is no significant difference in the gender-wise teacher's viewpoint towards factors responsible for low employability of management graduates.
- **H_{02}:** There is no significant difference in the institution-wise teacher's viewpoint towards factors responsible for low employability of management graduates.

3.5 RESEARCH METHODOLOGY

The nature of research design is exploratory-cum-descriptive in nature. Under the sample, data have been collected from the faculty members of different educational institutions, such as central universities, state universities, private universities, and IIMs. Under the area, Haryana, Punjab, and Chandigarh have been covered. Data have been collected from 160 respondents by comprising 80 male and 80 female teachers, under the gender-wise bifurcation and secondly under the institution-wise 40 teachers of central, state, private universities, and IIMs each. A pre-tested questionnaire was framed for the purpose of data collection. After the data collection to make the comparison between the views of male and female teachers t-test was applied and to analyze the institution-wise comparative viewpoint ANOVA was applied.

3.6 RESULTS AND DISCUSSIONS

3.6.1 GENDER-WISE TEACHERS' VIEWPOINT

Table 3.1 presents the gender-wise descriptive and inferential statistics of the teachers' viewpoint towards the factors responsible for low employability of management students. The male teachers focused highly on wrong selection procedure ($\bar{x}$=4.61, σ=0.618), lack of experience in real working environment ($\bar{x}$=4.57, σ=0.540), lack of research ($\bar{x}$=4.44, σ=0.673), lack of confidence ($\bar{x}$=4.40, σ=0.728) and perceptional differences between employer and fresher ($\bar{x}$=4.39, σ=0.637). On the other side, female teachers emphasized on lack of latest technological system in education ($\bar{x}$=4.77, σ=0.323), perceptional differences between employer and fresher ($\bar{x}$=4.70, σ=0.362), lack of experience in real working environment ($\bar{x}$=4.70, σ=0.412), old syllabus ($\bar{x}$=4.69, σ=0.416) and lack of confidence ($\bar{x}$=4.66, σ=0.424). Statistically, the results of t-test show that there is a significant difference in the viewpoint of male and female teachers of the selected institutions towards lack of confidence (p=0.000), old education system (p=0.000), lack of skilled education (p=0.001), perceptional differences between employer and fresher (p=0.000), old syllabus (p=0.000), lack of latest technological system in education (p=0.001), lack of communication in academic and industrial staff (p=0.001), wrong selection procedure (p=0.000), market image of educational institution (p=0.001) as factors responsible for low employability of management students at 5% level of significance, hence the null hypothesis (H01) is rejected.

TABLE 3.1 Gender-Wise Teachers' Viewpoint Towards Factors Responsible for Low Employability of Management Students

Statements	Males			Females			t-Test	
	N	$\bar{x}$	σ	N	$\bar{x}$	σ	t	Sig
Lack of confidence.	80	4.40	0.728	80	4.66	0.424	1.437	0.000*
Lack of guidance.	80	4.09	0.844	80	4.14	0.627	1.460	0.246
Lack of ambition.	80	4.37	0.809	80	4.39	0.601	1.162	0.272
Poor academic score.	80	4.15	0.883	80	4.31	0.731	2.314	0.491
Old education system.	80	4.33	0.710	80	4.61	0.490	1.992	0.000*
Lack of skilled education system.	80	4.21	0.765	80	4.58	0.497	2.736	0.001*
Perceptional differences between employer and fresher.	80	4.39	0.637	80	4.70	0.362	2.248	0.000*
Lack of participation in extra-curricular activities.	80	4.31	0.797	80	4.20	0.712	1.909	0.465
Old syllabus.	80	4.36	0.777	80	4.69	0.416	2.249	0.000*
Lack of latest technological system in education.	80	4.28	0.773	80	4.77	0.323	2.148	0.001*
Lack of experience in real working environment.	80	4.57	0.540	80	4.70	0.412	2.392	0.366
Lack of communication in academics and industrialists.	80	4.23	0.788	80	4.64	0.531	1.946	0.001*
Wrong selection procedure.	80	4.61	0.618	80	4.35	0.803	1.269	0.000*
Lack of skilled staff.	80	3.90	0.778	80	3.73	0.717	2.406	0.261
Lack of research.	80	4.44	0.673	80	4.63	0.534	3.853	0.066
Market image of educational institution.	80	3.93	0.846	80	4.42	0.521	2.527	0.001*
High salary demand.	80	4.18	0.829	80	4.08	0.702	1.843	0.300
Extra demand of security benefits.	80	4.10	0.850	80	3.90	0.597	2.767	0.079
Geographical factors.	80	4.15	0.755	80	3.94	0.669	2.913	0.058
Economic factors.	80	4.21	0.765	80	4.27	0.682	1.522	0.202

Note: N: Number of respondents; and *: Significant at 5% level.

Source: Primary survey (author's calculation).

3.6.2 *INSTITUTION-WISE TEACHERS' VIEWPOINT*

The institution-wise frequency distribution of teachers' viewpoint towards the reasons for low employability of management students indicates that most of the teachers of central universities strongly agree/agree with wrong selection procedure (n = 35, p = 87.50), lack of confidence (n = 34, p = 85), lack of experience in real working environment (n = 32, p = 80), lack of ambition (n = 31, p = 77.50), geographical factors (n = 30, p = 75), extra demand of security benefits (n = 30, p = 75), economic factors (n = 28, p = 70), old education system (n = 27, p = 67.50), perceptional differences between employer and fresher (n=26, p=65), lack of research (n = 26, p = 65), lack of latest technological system in education (n = 23, p = 57.50), old syllabus (n = 23, p = 57.50), market image of educational institution (n = 22, p = 55), lack of interest in multi-dimensional activities (n = 21, p = 52.50), high salary demand (n = 21, p = 52.50), lack of communication in academic and industrial staff (n = 19, p = 47.50), lack of skilled staff (n = 18, p = 45), lack of skilled education system (n = 17, p = 42.50) and poor academic score (n = 16, p = 40). Whereas, the teachers of state universities strongly agree/agree with lack of experience in real working environment (n = 49, p = 98), wrong selection procedure (n = 48, p = 96), lack of research (n = 46, p = 92), old syllabus (n = 45, p = 90), perceptional differences between employer and fresher (n = 45, p = 90), lack of confidence (n = 44, p = 88), lack of ambition (n = 42, p = 84), lack of interest in multi-dimensional activities (n = 40, p = 80), old education system (n = 40, p = 80), lack of education system (n = 38, p = 76), lack of latest technological system in education (n = 37, p = 74), lack of guidance (n = 36, p = 72), poor academic score (n = 36, p = 72), lack of communication in academic and industrial staff (n = 34, p = 68), high salary demand (n = 33, p = 66), economic factors (n = 33, p = 66), geographical factors (n = 31, p = 62) and extra demand of security benefits (n = 30, p = 60). On the other hand, the teachers of private universities strongly agree/agree with wrong selection procedure (n = 47, p = 94), lack of latest technological system in education (n = 46, p = 92), lack of confidence (n = 45, p = 90), old education system (n = 45, p = 90), lack of experience in real working environment (n = 44, p = 88), perceptional differences between employer and fresher (n = 42, p = 84), old syllabus (n = 41, p = 82), lack of communication in academic and industrial staff (n = 41, p = 82), lack of skilled education system (n = 40, p = 80), lack of research (n = 38, p = 76), market image of educational institution (n = 37, p = 74), lack of ambition (n = 36, p = 72), lack of guidance (n = 36, p = 72), economic factors (n = 35, p = 70), extra demand of security benefits (n = 34, p = 68), geographical factors (n = 32, p = 64), poor academic score (n = 32, p = 64) and high salary demand (n = 30, p = 60). However, the teachers of IIMs strongly agree/agree with lack of latest technological system in education (n = 27, p = 90), perceptional differences between employer and fresher (n = 26, p = 86.67), lack of experience in real working environment (n = 25, p = 83.33), lack of communication in academic and industrial staff (n = 25, p = 83.33), lack of research (n = 24, p = 80), old syllabus (n = 22, p = 73.33), poor academic score (n = 21, p = 70), lack of confidence (n = 20, p = 66.67), lack of skilled education system (n = 20, p = 66.67), old education system (n = 18, p = 60), lack of interest in multi-dimensional activities (n = 18, p = 60), market image of educational institution (n = 17, p = 56.67), lack of ambition (n = 17, p = 56.67), economic factors (n = 16, p = 53.33), high salary demand (n = 15, p = 50), lack of guidance (n = 15, p = 50), geographical factors (n = 13, p = 43.33), extra demand of security benefits (n = 12, p = 40). The overall results of frequency distribution of teachers' viewpoint towards the reasons for low employability of management students shows that most of the teachers strongly agree/agree with lack of experience in real working environment (n = 150, p = 88.24), lack of confidence (n = 143, p = 84.12), perceptional differences between employer and fresher (n = 139, p = 81.76), lack of research (n = 134, p = 78.82), lack of latest technological system in education (n = 133, p = 78.24), old syllabus (n = 131, p = 77.06), wrong selection procedure (n = 130, p = 76.47), old education system (n = 130, p = 76.47), lack of communication in academic and industrial staff (n = 119, p = 70), lack of skilled education system (n = 115, p = 67.65), lack of ambition (n = 114, p = 67.06), lack of interest in multi-dimensional activities (n = 106, p = 62.35), poor academic score (n = 105, p = 61.76), economic factors (n = 102, p = 60), market image of educational institution (n = 101, p = 59.41), lack of guidance (n = 81, p = 47.65), high salary demand (n = 80, p = 47.06) and geographical factors (n = 72, p = 42.35).

Table 3.2 indicates that teachers at Central Universities focused on wrong selection procedure ($\bar{x}$ = 4.52, σ = 0.569), lack of confidence ($\bar{x}$ = 4.45, σ = 0.529), lack of experience in real working environment ($\bar{x}$ = 4.40, σ = 0.435), lack of ambition ($\bar{x}$ = 4.39, σ = 0.619) and geographical factors ($\bar{x}$ = 4.36, σ = 0.512). Whereas, the teachers in State Universities emphasized on lack of experience in real working environment ($\bar{x}$ = 4.81, σ = 0.514), wrong selection procedure ($\bar{x}$ = 4.79, σ = 0.521), lack of research ($\bar{x}$ = 4.72, σ = 0.423), old syllabus ($\bar{x}$ = 4.59, σ = 0.465) and perceptional differences between employer and fresher ($\bar{x}$ = 4.55, σ = 0.467). On the other hand, teachers of Private Universities gave priority to wrong selection procedure ($\bar{x}$ = 4.71, σ = 0.420), lack of latest technological system in education ($\bar{x}$ = 4.64, σ = 0.469), lack of confidence ($\bar{x}$ = 4.61, σ = 0.487), old education system ($\bar{x}$ = 4.61, σ = 0.497) and lack of experience in real working environment ($\bar{x}$ = 4.59, σ = 0.521). However, the teachers of IIMs focused highly on lack of latest technological system in education ($\bar{x}$ = 4.91, σ = 0.316), perceptional differences between employer and fresher ($\bar{x}$ = 4.91, σ = 0.335), lack of experience in real working environment ($\bar{x}$ = 4.89, σ = 0.336), lack of communication in academic and industrial staff ($\bar{x}$ = 4.86, σ = 0.362) and lack of research ($\bar{x}$ = 4.84, σ = 0.399). As a whole, lack of confidence ($\bar{x}$ = 4.58, σ = 0.616) is the most important factor responsible for low employability of management students, followed by lack of guidance ($\bar{x}$ = 4.54, σ = 0.590), lack of ambition ($\bar{x}$ = 4.47, σ = 0.670), old education system ($\bar{x}$ = 4.41, σ = 0.553) and lack of research ($\bar{x}$ = 4.39, σ = 0.638). Statistically, ANOVA results show that there is a significant difference in the viewpoint of teachers of the selected institutions towards lack of guidance (p = 0.000), poor academic score (p = 0.001), lack of interest in multi-dimensional activities (p = 0.000), market image of educational institution (p = 0.000), extra demand of security benefits (p = 0.001) and economic factors (p = 0.002) at 5% level of significance hence the null hypothesis (H02) is rejected.

3.7 FACTOR ANALYSIS

To identify the factors responsible for low employability of management students, factor analysis is used, which is a statistical measure to reduce the initial variables as well as to combine together them for producing the broad range of factors on the basis of common variables (Inkova, 2014). The Cronbach's Alpha coefficient is 0.812, which presents the reliability and consistency of the data used in analysis (Nunnaly, 1978). To fulfill the minimum standard of factor analysis, two tests have been applied. Kaiser-Meyer-Olkin (KMO) examines the suitability of factor analysis and is a good measure of sampling adequacy. The calculated

TABLE 3.2 Institution-Wise Teachers' Viewpoint Towards Factors Responsible for Low Employability of Students

Statements	N	Central Universities		N	State Universities		N	Private Universities		N	IIMs		N	Total		ANOVA	
		$\bar{x}$	σ		$\bar{x}$	σ		$\bar{x}$	σ		$\bar{x}$	σ		$\bar{x}$	σ	F	Sign.
Lack of confidence.	40	4.45	0.529	40	4.52	0.646	40	4.61	0.487	40	4.73	0.450	160	4.58	0.616	1.532	0.212
Lack of guidance.	40	3.93	0.694	40	4.40	0.639	40	4.18	0.629	40	4.03	0.615	160	4.54	0.590	1.232	0.000*
Lack of ambition.	40	4.39	0.619	40	4.50	0.580	40	4.24	0.625	40	4.57	0.504	160	4.47	0.670	2.222	0.113
Poor academic score.	40	4.03	0.768	40	4.38	0.697	40	4.02	0.714	40	4.79	0.426	160	3.79	1.065	1.231	0.001*
Old education system.	40	4.30	0.648	40	4.44	0.577	40	4.61	0.497	40	4.67	0.486	160	4.41	0.553	2.253	0.346
Lack of skilled education system.	40	4.07	0.730	40	4.42	0.609	40	4.52	0.505	40	4.67	0.479	160	4.37	0.669	3.223	0.172
Perceptional differences between employer and fresher.	40	4.30	0.687	40	4.55	0.467	40	4.58	0.499	40	4.91	0.335	160	4.35	0.655	1.223	0.164
Lack of participation in extra-curricular activities.	40	4.17	0.747	40	4.50	0.647	40	3.96	0.727	40	4.60	0.498	160	3.86	1.066	3.768	0.000*
Old syllabus.	40	4.23	0.768	40	4.59	0.465	40	4.58	0.575	40	4.83	0.379	160	4.30	0.883	1.325	1.218
Lack of latest technological system in education.	40	4.23	0.698	40	4.40	0.670	40	4.64	0.469	40	4.91	0.316	160	4.14	0.948	2.258	0.464
Lack of experience in real working environment.	40	4.40	0.435	40	4.81	0.514	40	4.59	0.521	40	4.89	0.336	160	4.35	0.692	1.537	0.286
Lack of communication in academics and industrialist.	40	4.13	0.757	40	4.36	0.693	40	4.56	0.501	40	4.86	0.362	160	4.31	0.722	2.786	0.282
Wrong selection procedure.	40	4.52	0.569	40	4.79	0.521	40	4.71	0.420	40	3.50	0.630	160	4.28	0.912	1.251	0.226
Lack of skilled staff.	40	4.10	0.672	40	3.82	0.748	40	3.82	0.720	40	3.57	0.679	160	4.31	0.850	1.221	0.184
Lack of research.	40	4.27	0.784	40	4.72	0.423	40	4.48	0.580	40	4.84	0.399	160	4.39	0.638	2.523	0.146

TABLE 3.2 *(Continued)*

Statements	N	Central Universities		N	State Universities		N	Private Universities		N	IIMs		N	Total		ANOVA	
		$\bar{x}$	σ		$\bar{x}$	σ		$\bar{x}$	σ		$\bar{x}$	σ		$\bar{x}$	σ	F	Sign.
Market image of educational institution.	40	4.17	0.712	40	3.82	0.825	40	4.34	0.479	40	4.60	0.500	160	4.33	0.972	1.232	0.000*
High salary demand.	40	4.15	0.802	40	4.34	0.717	40	4.00	0.670	40	4.13	0.730	160	3.72	1.109	2.567	0.241
Extra demand of security benefits.	40	4.35	0.770	40	4.00	0.756	40	4.04	0.533	40	3.63	0.615	160	3.90	1.018	2.786	0.001*
Geographical factors.	40	4.36	0.512	40	4.08	0.695	40	4.02	0.685	40	3.77	0.626	160	4.34	0.943	1.256	0.121
Economic factors.	40	4.33	0.656	40	4.18	0.720	40	4.10	0.707	40	4.57	0.538	160	3.86	1.054	3.325	0.002*

Note: N: Number of respondents; *: Significant at 5% level.

Source: Primary survey (author's calculation).

value of KMO is 0.762, which presents the appropriateness of factor analysis (Table 3.3).

TABLE 3.3 Coefficients of Sampling Adequacy

Coefficients		**Value**
Kaiser-Meyer-Olkin (KMO)		0.762
Bartlett's test of sphericity	Chi-square	4466.260
	d.f.	190
	Sig.	0.000*

Note: N: Number of respondents; and *: Significant at 5% level.

Source: Compiled from primary data (SPSS output).

For the analysis of Anti-image matrix, diagonals contained the value more than 0.6 are retained. Measures of sampling adequacy (MSA) are presented through the diagonal elements of anti-image correlation matrix. MSA facilitates to achieve more stable results of factor analysis and to drop out the collections of variables which seems to be non-factorable (Pathak, 2014).

3.7.1 SCREE PLOT

Scree plot is a graphical presentation of factors to be retained in factor analysis. The factors that lie above the point of inflection must be retained in factor analysis (Cattell, 1966). Scree plot exhibits that top four factors have Eigen values greater than 1. Thus, the graphical output under Scree plot also support the statistical output of PCA and confirms the extraction of four factors under factor analysis.

The factors were drawn through principal component method that reduces a large number of variables with preserving much information of initial variables. The 20 variables were divided into the four factors. The initial Eigen values represent the total dispersion attributed to that factor (Figure 3.1).

The variance share of first factor is 7.839 (22.56%) of the total variances, second factor is 3.294 (21.75%) of the total variances, third factor is 2.896 (21.66%) of the total variances and likewise fourth factor is 2.418 (16.27%) of the total variances. The first four factors combined accounts

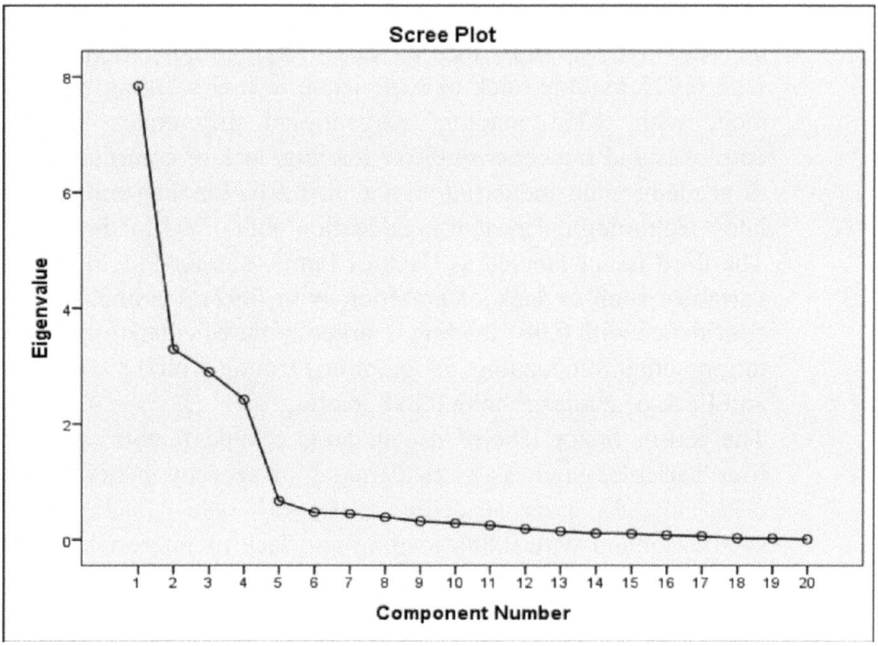

FIGURE 3.1 Scree plot of the variables.

Source: Compiled from primary data (SPSS output).

for 82.24% of the total variance. The reliability of above four factors was measured by Cronbach's Alpha coefficient that has value of 0.839, 0.953, 0.952, and 0.915, respectively.

Table 3.6 indicates the communalities that amount to dispersion of variables shared by four factors. The values of communalities vary from 0.629 to 0.937 and all the values were more than 0.5 as required. Therefore, all the variables were retained, which shows the adequate explanation of the drawn factors.

Table 3.7 exhibits the factor labeling known as rotated component matrix. This matrix presents the association between 20 variables formed four factors:

- The first factor labeled as 'Government policies factors' includes six variables such as old education system with 0.929 loading, economic factors with 0.893 loading, lack of research with 0.881 loading, lack of skilled staff with 0.798 loading, old syllabus with 0.771 loading and wrong selection procedure with 0.744 loading.

- The second factor labeled as 'Educational institutional factors' includes five variables such as lack of skilled education system with 0.933 loading, lack of experience in real working environment with 0.931 loading, perceptional differences between employer and fresher with 0.920 loading, lack of communication in academic and industrial staff with 0.916 loading and lack of latest technological system in education with 0.774 loading.
- The third factor labeled as 'lack of family support' includes five variables such as lack of ambition with 0.921 loading, lack of confidence with 0.901 loading, market image of educational institution with 0.896 loading, geographical factors with 0.895 loading and Lack of guidance with 0.888 loading.
- The fourth factor labeled as 'students related factors' includes four variables such as extra demand of security benefits with 0.937 loading, poor academic score with 0.900 loading, high salary demand with 0.886 loading and lack of interest in multidimensional activities with 0.787 loading (Tables 3.4–3.7).

3.8 IMPLICATIONS OF STUDY

The study represents the factors responsible for low employability of management students as per teacher's viewpoint of different universities. By the consideration of such factors, educational institutions can find out that where the lack exists on their behalf and after that what steps can be initiated to solve the issues. As well as students can also analyze that where they are responsible for not getting the job and what efforts they can make to remove the barriers that create the situation of unemployability. Neither only students nor only educational institutions are responsible for the low employability of management graduates, both are equally responsible. By the consideration of results of the present study both students and educational institutions find their faults and take necessary measures to cope with such kind of problems.

3.9 RESEARCH GAPS AND USEFULNESS OF THE STUDY

The present study reveals those factors which are highly responsible for low employability of management students, while in the previous study focus have been made on the engineering students only. In the current

TABLE 3.4 Anti-Image Matrix (Measure of Sampling Adequacy)

Anti-Image	1	2	3	4	5	6	7	8	9	10	11	12	13	14	15	16	17	18	19	20
1	**0.609**[a]	−0.827	−0.620	−0.226	−0.282	−0.503	0.568	0.038	0.057	0.287	0.382	−0.322	−0.237	0.436	−0.032	0.233	0.229	0.224	−0.220	−0.250
2	−0.827	**0.663**[a]	0.283	0.256	0.293	0.238	−0.502	−0.079	−0.097	−0.266	−0.224	0.325	0.035	−0.472	0.228	−0.236	−0.263	−0.089	0.236	0.092
3	−0.620	0.283	**0.772**[a]	0.207	0.222	0.483	−0.289	−0.052	−0.004	−0.270	−0.458	0.263	0.206	−0.258	−0.297	−0.354	−0.204	−0.203	0.259	0.230
4	−0.226	0.256	0.207	**0.547**[a]	0.330	0.262	−0.325	−0.229	−0.277	−0.264	−0.226	0.222	0.056	−0.430	0.345	−0.348	−0.228	−0.726	0.347	−0.226
5	−0.282	0.293	0.222	0.330	**0.786**[a]	0.230	−0.274	−0.284	−0.469	0.048	−0.222	0.225	−0.352	−0.263	−0.220	−0.239	−0.056	−0.232	0.224	−0.468
6	−0.503	0.238	0.483	0.262	0.230	**0.656**[a]	−0.560	−0.002	−0.049	0.243	−0.922	0.062	0.238	−0.230	−0.054	−0.420	−0.229	−0.232	0.449	−0.002
7	0.568	−0.502	−0.289	−0.325	−0.274	−0.560	**0.683**[a]	0.228	0.094	−0.058	0.353	−0.349	−0.078	0.283	−0.270	0.357	0.075	0.202	−0.389	0.063
8	0.038	−0.079	−0.052	−0.229	−0.284	−0.002	0.228	**0.832**[a]	0.208	−0.222	0.026	−0.228	0.083	0.236	−0.254	0.282	−0.258	−0.097	−0.287	0.085
9	0.057	−0.097	−0.004	−0.277	−0.469	−0.049	0.094	0.208	**0.870**[a]	0.083	0.058	−0.247	0.029	0.240	−0.085	0.228	0.023	0.202	−0.243	0.035
10	0.287	−0.266	−0.270	−0.264	0.048	0.243	−0.058	−0.222	0.083	**0.864**[a]	−0.257	−0.403	−0.047	0.275	−0.226	0.025	0.092	0.222	−0.027	−0.225
11	0.382	−0.224	−0.458	−0.226	−0.222	−0.922	0.353	0.026	0.058	−0.257	**0.684**[a]	−0.220	−0.238	0.277	0.222	0.479	0.254	0.066	−0.504	−0.028
12	−0.322	0.325	0.263	0.222	0.225	0.062	−0.349	−0.228	−0.247	−0.403	−0.220	**0.772**[a]	0.025	−0.252	0.088	−0.447	−0.252	−0.009	0.458	0.096
13	−0.237	0.035	0.206	0.056	−0.352	0.238	−0.078	0.083	0.029	−0.047	−0.238	0.025	**0.932**[a]	−0.045	0.022	0.045	−0.222	−0.052	−0.028	−0.022
14	0.436	−0.472	−0.258	−0.430	−0.263	−0.230	0.283	0.236	0.240	0.275	0.277	−0.252	−0.045	**0.676**[a]	−0.497	0.234	0.286	0.279	−0.254	−0.072
15	−0.032	0.228	−0.297	0.345	−0.220	−0.054	−0.270	−0.254	−0.085	−0.226	0.222	0.088	0.022	−0.497	**0.830**[a]	−0.032	−0.022	−0.242	0.063	−0.259
16	0.233	−0.236	−0.354	−0.348	−0.239	−0.420	0.357	0.282	0.228	0.025	0.479	−0.447	0.045	0.234	−0.032	**0.628**[a]	0.282	0.232	−0.922	0.036
17	0.229	−0.263	−0.204	−0.228	−0.056	−0.229	0.075	−0.258	0.023	0.092	0.254	−0.252	−0.222	0.286	−0.022	0.282	**0.802**[a]	−0.353	−0.236	0.005
18	0.224	−0.089	−0.203	−0.726	−0.232	−0.232	0.202	−0.097	0.202	0.222	0.066	−0.009	−0.052	0.279	−0.242	0.232	−0.353	**0.698**[a]	−0.227	0.262
19	−0.220	0.236	0.259	0.347	0.224	0.449	−0.389	−0.287	−0.243	−0.027	−0.504	0.458	−0.028	−0.254	0.063	−0.922	−0.236	−0.227	**0.622**[a]	−0.027
20	−0.250	0.092	0.230	−0.226	−0.468	−0.002	0.063	0.085	0.035	−0.225	−0.028	0.096	−0.022	−0.072	−0.259	0.036	0.005	0.262	−0.027	**0.876**[a]

[a]Response retained.
Source: Compiled from primary data (SPSS output).

TABLE 3.5 Extraction Sums of Squared Loadings and Rotation Sums of Squared Loadings

Components	Initial Eigen Values			Extraction Sum of Squared Loadings			Rotation Sum of Squared Loadings		
	Total	Percentage of Total Variance	Cumulative (Percent)	Total	Percentage of Total Variance	Cumulative (Percent)	Total	Percentage of Total Variance	Cumulative (Percent)
1	7.839	39.195	39.195	7.839	39.195	39.195	4.512	22.558	22.558
2	3.294	16.470	55.565	3.294	16.470	55.665	4.349	21.746	44.304
3	2.896	14.481	70.146	2.896	14.481	70.146	4.332	21.660	65.964
4	2.418	12.089	82.235	2.418	12.089	82.235	3.254	16.271	82.235
5	0.671	3.357	85.592						
6	0.475	2.374	87.966						
7	0.448	2.240	90.206						
8	0.388	1.941	92.147						
9	0.315	1.577	93.724						
10	0.281	1.406	95.130						
11	0.247	1.236	96.366						
12	0.184	0.919	97.285						
13	0.143	0.716	98.002						
14	0.107	0.535	98.536						
15	0.101	0.503	99.039						
16	0.078	0.391	99.430						
17	0.061	0.304	99.734						
18	0.025	0.123	99.856						
19	0.023	0.114	99.971						
20	0.006	0.029	100.000						

Source: Compiled from primary data (SPSS output).

TABLE 3.6 Extraction of the Variables

Sl. No.	Statements	Initial	Extraction
1.	Lack of confidence	1.000	0.867
2.	Lack of guidance	1.000	0.827
3.	Lack of ambition	1.000	0.929
4.	Poor academic score	1.000	0.821
5.	Old education system	1.000	0.909
6.	Lack of skilled education system	1.000	0.933
7.	Perceptional differences between employer and fresher	1.000	0.897
8.	Lack of interest in multi-dimensional activities	1.000	0.709
9.	Old syllabus	1.000	0.629
10.	Lack of latest technological system in education	1.000	0.705
11.	Lack of experience in real working environment	1.000	0.937
12.	Lack of communication in academic and industrial staff	1.000	0.908
13.	Wrong selection procedure	1.000	0.642
14.	Lack of skilled staff	1.000	0.696
15.	Lack of research	1.000	0.806
16.	Market image of educational institution	1.000	0.852
17.	High salary demand	1.000	0.806
18.	Extra demand of security benefits	1.000	0.902
19.	Geographical factors	1.000	0.842
20.	Economic factors	1.000	0.829

Source: Compiled from primary data (SPSS output).

TABLE 3.7 Factors Labeling and Loading

Factor(s)	Factors Labeling	Statements	Factor Loading			
			I	II	III	IV
Factor I	Government policies factors	Old education system.	**0.929**			
		Economic factors.	0.893			
		Lack of research.	0.881			
		Lack of skilled staff.	0.798			
		Old syllabus.	0.771			
		Wrong selection procedure.	0.744			
Factor II	Educational institutional factors	Lack of skilled education.		0.933		
		Lack of experience in real working environment.		0.931		
		Perceptional differences between employer and fresher.		0.920		

TABLE 3.7 *(Continued)*

Factor(s)	Factors Labeling	Statements	Factor Loading			
			I	II	III	IV
		Absence of academic-industry interface.		0.916		
		lack of latest technological system in education		0.774		
Factor III	Lack of family support	Lack of ambition.			0.921	
		Lack of confidence.			0.901	
		Market image of educational institution.			0.896	
		Geographical factors.			0.895	
		Lack of guidance.			0.888	
Factor IV	Students related factors	Extra demand of security benefits.				**0.937**
		Poor academic score.				0.900
		High salary demand.				0.886
		Lack of interest in multi-dimensional activities.				0.787
Value of Cronbach's Alpha Coefficient of the Factors						
0.839	0.953	0.952	0.915			

Source: Compiled from primary data (SPSS output).

research data have collected from those teachers who teach the management graduates only. Under the study bifurcation has been made on gender-wise and institution-basis. It represents the views of male and female teachers separately as well as IIMs, central, state, and private universities' teachers separately, which differentiate the present study from previous conducted studies. So, this study is very useful for the management students and all educational institutions. As well, study will provide reference to the researchers to research in the same stream.

3.10 LIMITATIONS AND FURTHER SCOPE OF RESEARCH

The conducted study is restricted to area of Haryana, Punjab, and Chandigarh. So, the results cannot be applied to the whole education system. Due to the small sample size, the outcomes of the study cannot be generalized. The study has been conducted on gender and institution-basis, future it may be carried out based on designation and age, etc., by using the different statistical tools such as correlation and regression, etc.

KEYWORDS

- **education system**
- **educational institution**
- **employability**
- **gender**
- **geographical factors**
- **holistic management education**
- **matrix**
- **scree plot**

REFERENCES

Aida, B., & Norailis, A., et al., (2015). Critical success factors of graduate employability programs. Journal of Economics, Business and Management, 3(8), 767–771.

Bala, R., & Singh, S., (2020). Employability skills of management students: A study of banker's viewpoint. Psychology and Education Journal, 57(9), 2908–2914.

Bala, R., & Singh, S., (2020). Perception of students towards employability skills. Shodh Sarita, 7(28), 107–112.

Bala, R., & Singh, S., (2021). Employability skills of management students: A study of teacher's viewpoint. Material Today: Proceedings.

Behle, H., (2020). Students' and graduates' employability. A framework to classify and measure employability gain. Policy Reviews in Higher Education, 4(1), 105–130.

Cheong, K., Hill, C., et al., (2016). Employing the 'unemployable': Employers perception of Malaysian graduates. Studies in Higher Education, 41(12), 2253–2270.

Curtis, S., & Shani, N., (2002). The effect of taking paid employment during term-time on students' academic studies. Journal of Further and Higher Education, 26(2), 129–138.

Frankham, J., (2017). Employability and higher education: The follies of the 'productivity challenge' in the teaching excellence framework. Journal of Education Policy, 32(5), 628–641.

Jackson, D., (2013). Student perceptions of the importance of employability skill provision in business undergraduate programs. Journal of Education for Business, 88(5), 271–279.

Jenssen, (2011). Background and employability of mobile vs. non-mobile students. Tertiary Education and Management, 17(2), 79–100.

Jill, M., (2009). Working conditions in hospitality: Employees' views of the dissatisfactory hygiene factors. Journal of Quality Assurance in Hospitality & Tourism, 10(1), 23–43.

Leach, T., (2017). 'It seems at the moment my career is dependent on factors outside of my control: 'Reflections on graduates' experiences of employment and career enactment in an era of economic uncertainty and austerity. British Journal of Guidance & Counselling, 45(2), 188–198.

Moira, C., Siobhan, L., & Rosemary, S., (2018). Academic success: Is it just about the grades? Higher Education Pedagogies, 3(1), 434–439.

Paadi, K., (2014). Perceptions of employability skills necessary to enhance human resource management graduates prospects of securing a relevant place in the labor market. European Scientific Journal, 129–143.

Pan, Y. J., & Lee, L. S., (2011). Academic Performance and Perceived Employability of Graduate Students in Business and Management-An Analysis of Nationwide Graduate Destination Survey (Vol. 25, pp. 91–103). Paper presented at the conference on International Conference on Asia Pacific Business Innovation & Technology Management. Available at www.sciencedirect.com (accessed on 20 February 2024).

Pawan, S., Budhwar, S., & Bhatnagar, J., (2005). Women in management in the new economic environment: The case of India. Asia Pacific Business Review, 11(2), 179–193.

Richardson, S., & Butler, G., (2012). Attitudes of Malaysian tourism and hospitality students' towards a career in the industry. Asia Pacific Journal of Tourism Research, 17(3), 262–276.

Saunders, V., & Zuzel, K., (2010). Evaluating employability skills: Employer and student perceptions. Bioscience Education, 15(1), 1–15.

Sin, C., Tavares, O., & Amaral, A., (2016). Who is responsible for employability? student perceptions and practices. Tertiary Education and Management, 22(1), 65–81.

Speight, S., Lackovic, N., & Cooker, L., (2013). The contested curriculum: Academic learning and employability in higher education. Tertiary Education and Management, 19(2), 112–126.

Succi, C., & Canovi, M., (2020). Soft skills to enhance graduate employability: Comparing students and employers' perceptions. Studies in Higher Education, 45(9), 1834–1847.

Tymon, A., (2013). The student perspective on employability. Studies in Higher Education, 38(6), 841–856.

William, E., Donald, Y., & Melanie, A., (2019). The undergraduate self-perception of employability: Human capital, careers advice, and career ownership. Studies in Higher Education, 44(4), 599–614.

Yorke, M., & Thomas, L., (2003). Improving the retention of students from lower socio-economic groups. Journal of Higher Education Policy and Management, 25(1), 63–74.

Zahid, M. M., Naeem, A., et al., (2013). Business education curriculum and management skills: A viewpoint of students and managers. Academic Research International Journal, 4(4), 431–441.

CHAPTER 4

Heading Toward the Solution of Employability Issues Through Interdisciplinary Management Education

RUBY SHARMA,[1] PALLAVI PAHUJA,[1] GARIMA SINGH,[1] and ANSHUMAN SHARMA[2]

[1]*Chandigarh Business School of Administration, Punjab, India*

[2]*Assistant Professor, College of Business Administration, Ajman University, Dubai, UAE*

ABSTRACT

Employability issues among management students have been a topic of research as well as discussion since long. Various studies and research have been conducted with every research suggesting remedial action along with industry personnel coming forward with their requirements. The students aim for global companies to recruit them. The companies, on the other hand, are looking for competent and skilled workforce but both ends are hardly able to lessen the gap. Why is it so? Where does the problem lie? Is it with the management institutions? Is it the inability of students? Or is there a lack of understanding of industry's requirements among the institutions? This question is what the chapter aims to address.

The present chapter begins with explaining the concept of employability from different view-points and its need while also stating the skills generally agreed upon which are required to be considered employable with the

Interdisciplinary Approaches in Management Education.
Tanusree Chakraborty, Ramandeep Saini, Rajiv Khosla, Nandita Mishra, and Ashok Natarajan (Eds.)

backdrop of management field. Laying the foundation, the chapter then proceeds to address the lack of factors that have resulted in employability issues prevailing among management students. Interestingly, it was seen that it's not a single factor but a combination of a number of factors which resulted in cropping up of employability issues such as lack of soft skills, lack of career advice, etc., while also addressing the managers' point of view regarding the same. Further, the chapter highlights the reasons why employability issue has become prominent among management graduates with finally suggesting actions or remedies that can be taken up to tackle these issues.

4.1 INTRODUCTION

Amidst the thousands of management graduates churned out by B-schools across the country, barely 7% turn out to be employable (ASSOCHAM, 2016). Why is the situation so discouraging? Let's try to investigate why things are going wrong.

In 1991, liberalization era changed everything and opened up ample of job prospects for the Indian youth, especially in the field of management. Indian institute of management (IIM) became the center of attraction when it came to campus hiring and with the passage of time the demand for managers grew exponentially. This led to emergence of several B-schools offering PGDM and addition of management as a subject in other institutions. The advent of technology not only revolutionized the industries but also changed the way the businesses functioned (Jetter, Satzger, & Neus, 2009). This led to a drastic change in skills demanded by top corporate leaders. The mere knowledge of static subjective theories and tools no longer proved to be sufficient for this dynamic technological era. Business schools have been one of the first to adopt the realities of post liberalization changes (Narayanan Ramaswamy, Partner, Education, and Skill development sector, KPMG). Even then, a significant gap got created between academics and skills. This is where the problem began and resulted in creation of a gap between what is taught and what is required in the market, commonly termed as employability gap.

The issue of employability stems not only due to infrastructural incapability but also due to a peculiar predisposition among students that management degree is ticket to high paying corporate world. Large section of students considers this degree as mere formality rather than exploring its offerings and widening their horizon. Furthermore, they also ignore the power of soft skills and aptitude, which are the fundamental aspect of every job. Businesses are in search of those graduates who can not only plan proactively but also ensure the adaptability of their plan in dynamic environment.

But sadly, the reality is dictating something else. For instance, the eighth edition of India Skills Report (ISR) released in 2021 found that only 46.59% of MBA graduates are employable, which represents the grim situation in contrast to us being proud of having highest workforce. According to companies, the management graduates lack in communication skills, which means that the graduates are not able to demonstrate their internships, projects, volunteer work with that much clarity and conviction.

The graduates, institutes, and B-schools should accept that the expectations of employers and corporate leaders have undergone a drastic change, and the stakeholders should not only try to make the educational content specific but also ensure that it's focused on soft skills and analytical learning. The students should be made aware about the latest trends and technologies too. Above all, it should prompt them to explore and develop a questioning spirit.

4.2 HIGHLIGHTS

With this chapter, the aim is to discuss the employability issues among management graduates, potential reasons and what can be done to overcome these issues. The present chapter moves along the lines of work done over the years and gradually shifts to present scenario while highlighting the following aspects:

- The meaning of employability over the years by researchers. Further, the necessity of being employable is also explored.
- Some of the desired employability skills in reference to existing research.
- What is the present scenario of employability issues among graduates and traversing through some of the potential reasons?
- Finally, what are the some of the remedies that can be undertaken by various shareholders to overcome these issues?

4.3 BACKGROUND

4.3.1 EMPLOYABILITY

- Hillage & Pollard (1998) defined employability as 'the capability to move self-sufficiently within the labor market to realize potential through sustainable employment.
- Harvey (2001) described employability is the art of a graduate to get a decent job. It is not the set of skills, but a bunch of experiences and attitudes developed during higher level learning, thus, is not a "product" but an act of transformational learning.
- Yorke (2004) mentioned that employability is defined as a set of achievements comprising of skills, understanding, and personal attributes that make graduates more likely to gain employment and be successful in their chosen occupation, and which benefits them, the workforce, the community as well as the economy.
- Heijden et al. (2005) stated that the ability to obtain new employment, if required that is to be independent in labor market by being willing and able to manage their own employment transitions between and within organizations.
- Berntson (2008) defined employability as an individual's perception of his/her possibilities of obtaining attractive employment opportunities. Employability is distinctive as – actual employability (objective employability) and perceived employability (subjective employability).

4.3.2 WHY IS IT NECESSARY?

Employability is all about getting and sustaining a job, while constantly upgrading your skillset to adapt to the changing nature of your present job as well as preparing for future prospects According to Kempster H., following justifies the necessity of employability skills:

1. **Remaining Competitive:** When you constantly upgrade your skills, it helps you not only to get a job and keep it but also prepares you for future prospects.
2. **Achieving Your Goals:** Your career involves different job roles and employers, even the current job you are in may change in

nature over time. Thus, employability skills are useful as they are transferable and will help you in achieving your goals.

3. **Enhancing Your Education:** Learning through a variety of methods such as volunteering, academic studies and work experience helps in development of different aspects of personality.
4. **Dynamic World:** The world is changing continuously, thanks to technology. It has led to change in demands of corporates too. To keep up with these changes, it is essential to upgrade your skill.
5. **A Degree is Not Enough:** In these competitive times, even the students of prestigious institutions lack job opportunities due to lack of skills.

4.3.3 EXAMPLES OF EMPLOYABILITY SKILLS

Enhancing employability skills in management education (ME) is considered as an important task by all universities and colleges with the skills required by management graduates. Employers prefer flexible and diligent individuals, the ones with subject knowledge and ability to negotiate (Saunders & Zuzel, 2010). The employability skills include not only hard skills but following as well:

1. **Communication:** It is the backbone of management. It not only helps in coordination and facilitates decision making but also helps in efficient planning, smooth and efficient working of an organization. It strengthens managerial efficiency, promotes co-operation and healthy spirit among employees. Leadership seems empty without effective communication.
2. **Leadership:** It is one of the core functions of management. It increases the efficiency and helps to achieve organizational goals. Leadership is an umbrella term that includes factors like – motivating the team, building confidence, building a conducive environment for work, and building the morale of the team.
3. **Organization Skills:** Imagining a single day in corporate life without time management, prioritizing events, multi-tasking is very difficult. These organization skills not only boost the productivity, but also make you an efficient employee.

4. **Adaptation:** Corporate world is extremely dynamic and the trends, goals change very fast. Adaptation is a have-to-have thing here. You cannot stick to a method of working that was used years ago. One needs to be fast in adapting the managerial skills as per the latest requirements and trends of the industries.
5. **Tech Savvy:** In current scenario, the organizations have employees from different age-groups, each of them trying to do justice to the specific role he is given and together they drive the voyage of the organization toward success. Although, sometimes it becomes imperative to introduce latest technologies to the employees that has the potential to transform the work and can give more productive results. Hence, the managers need to be tech-savvy, not tech-shy. The technological world is changing at a very fast pace. In order to work in these forever expanding IT disruptions, it becomes the responsibility of the managers to adapt and then help the employees to adapt. Making them tech-savvy is the best thing to start with.
6. **Problem Solving:** Organizations are complex systems and problems can arise at any point of time, the set of activities that help to sort-out the problem is called problem-solving. The capacity of a manager to find solutions to problems defines his ability to have good problem-solving skills. It is a process that can make a manager more proficient at his job. Right from understanding the problem to devising, evaluating, and implementing the solution; a manager must be very good in all the aspects.
7. **Self-Awareness:** It is something that helps you understand yourself and others. It tells a lot about yourself, about how others see you, your psychological temperament. Self-awareness is a mirror of the way you respond to others in real-time. When a manager is able to see himself clearly, it helps him make better/informed decisions, develop significant and meaningful relationship. A manager could lead others in an effective way. The more you know yourself, the better you can manage others, the better you can perform the duties assigned.
8. **Emotional Intelligence:** Goleman (2011) while telling the Harvard Business Review about the importance of emotional

intelligence in leadership, stated, "The most effective leaders are all alike in one crucial way."

The ability to understand, handles, and manage our emotions and recognizing and influencing the emotions of people around us is called emotional intelligence (Mayer & Salovey, 1990).

It includes following competencies:

1. **Self-Awareness:** It is the core layer of emotional intelligence. It emphasizes not only on the ability to be aware of your own strengths and weaknesses, but also covers aspects like recognizing your emotions. It makes you understand the effect emotions have on you and how it influences your team's performance.
2. **Self-Management:** It is the ability to manage and channelize your emotions specifically in stressful instances when things are not going as per plan. It is about maintaining a positive attitude irrespective of the situation.
3. **Social Awareness:** It is indeed very important to understand your emotions, but that is not enough always. As a manager, you must know the art of reading a room. Social awareness is the ability to understand others' emotions and other dynamics within the organization. Managers must practice empathy. The ability to understand colleagues' emotions and perspectives, which enable the manager to communicate effectively with employees.
4. **Relationship Management:** It is the ability to influence, mentor other employees and resolve conflicts that arise. Some managers prefer avoiding conflicts as much as they can, but it is very important to address the issues immediately and properly. Sometimes avoiding a conflict or ignoring it make the situation worse.

Managers set the bars and goals of their respective departments. If they don't have emotional intelligence, it could result to a situations like lower employee engagement and a high attrition rate. The consequences are not healthy for the organization.
The wheel shown in Figure 4.1 highlights the essential employability skills.

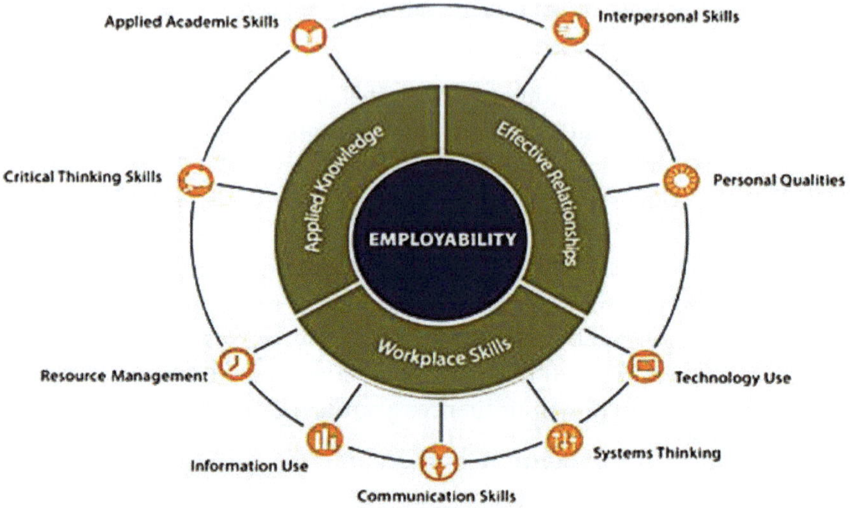

FIGURE 4.1 Employability skills.

Source: https://cte.ed.gov/initiatives/employability-skills-framework.

4.3.4 HOW TO DEVELOP EMPLOYABILITY SKILLS?

Various research has been conducted to identify the desired skills required for graduate employment and considerable agreement have been found among employers, irrespective of nature of job, on the skills valued most (Vanitha & Jaganathan, 2018). The key to developing employability skills is being aware of the changes in business environment. However, awareness alone would not suffice, so following actions should be taken:

1. **Emphasize on Soft Skills:** Today, the businesses are rapidly transforming around soft skills such as teamwork, time management, etc. Hence, employers are emphasizing more on these skills. To develop these skills, volunteering, community projects, etc., are the way to go.
2. **Networking:** Professional relationships help develop one's career and attain certain goals.
3. **Put Yourself in Employer's Shoes:** To get noticed by the prospective employer, you need to understand their needs. For this you need to understand your core skills and demonstrate them.

4. **Social Media:** In today's world, employers use social media platforms such as LinkedIn, Facebook, and Twitter, etc. So, it is up to students to make sure to put their best foot forward to get noticed.
5. **Prepare Smartly:** To prove yourself as a prospective candidate, you should be well versed with the company you are interviewing for.

4.4 EMPLOYABILITY ISSUES

Employability statistics for management have continued to pay a disheartening picture. There has been a decline in employability among management graduates with current being 46.59% as opposed to 54% in 2020 (India Skill Report, 2021). Why is it so? What are the issues? Well, it turns out that over the time several issues cropped up and contributed to this decline:

1. **Lack of Academic-Industry Interface:** In a rush to implement western curriculum in some way or the other, the Indian Management Institutes are lacking in developing an Indian Business environment friendly curriculum. The current curriculum is generic in nature and thus does not account for dynamic nature of market. Moreover, the curriculum doesn't go in-line with the market trends. This creates an academic-industry gap resulting in poor employability skills.
2. **Lack of Soft Skills:** Majority of the students focus on acquiring hard skills, i.e., subjective knowledge only while not working on soft skills such as communication skills, technical skills, etc. Employers have time and again pointed out that majority of management graduates lack soft skills with the foremost being communication skills followed by critical thinking skills, technology skills, real-world experience, and teamwork.
3. **Lack of Work Experience:** Majority of students joining management courses in India do not have work experience in contrast to the west, where a certain period of work experience is a pre-requisite. This results in students not being able to link the study material with real life scenarios, thus creating a gap. Having a prior work experience helps in better understanding and exploring the course work.

4. **Management Degree:** A status symbol-Even today, students view a management degree more of a status symbol rather than a means of expanding their knowledge and skills. It is seen as an easy means to enter the corporate world without considering their existing skill set. The students thus are more focused on packages and multinational corporations rather than developing their employability skills.

5. **Lack of Adequate Career Advice:** Whilst most of the universities offer career support, the advice for these career centers is usually generic and is not tailored to each individual student's needs. Even the students don't understand the need for performing SWOT (strength, weakness, opportunities, strengths) analysis to know where they stand in terms of skills. Furthermore, many students do not visit or consult the career center during their studies believing that higher CGPA is more than enough. Due to this majority of students end up with no placements with some even being in debt.

The People Management Report (2019) conducted a survey of almost 3,000 employees to know what the areas are where their managers lack. The graphical representation is shown in Figure 4.2.

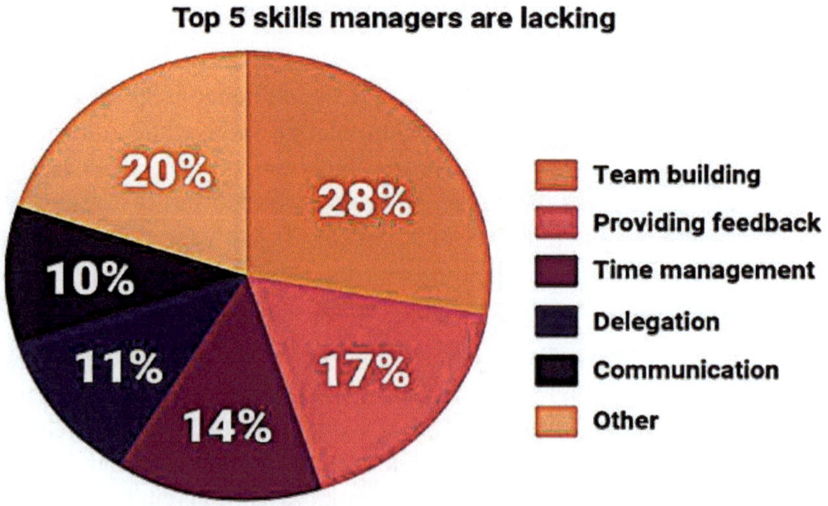

FIGURE 4.2 Issues with managers.

Source: Survey of People Management Report (2019).

The major areas where they feel managers lack are:

- Team building;
- Providing feedback;
- Time management;
- Delegation;
- Communication.

4.4.1 REASONS BEHIND EMPLOYABILITY ISSUES

Every year lakhs of students enroll in management students with hopes of landing a high paying corporate job. In reality, only a few are able to land their dream jobs while majority are left with struggles. The lack of employability among students has nothing but increased in past few years. Time and again, companies have pointed out the skills they have been looking for, yet students are lacking in those areas. We have already discussed the issues but the question that arises is why so such issues exist? The several factors that contribute to the cropping up of these issues are as follows:

1. **Increase in Number of Institutes:** Over the time, the demand for the management graduates has just increased. With this the number of institutes offering management degrees too has risen at commensurate rate. Though all of the institutions barring the top IIMs promise of great infrastructural and educational capabilities, they still lack in a basic area, i.e., quality control. Most of the institutions are mainly concerned with filling up of seats and getting a return on investment as soon as possible. Many such institutions are either unable to generate sufficient placements or don't even have placement cell of their own and market themselves falsely. As a result of this, every year thousands of students suffer since they are not getting enough opportunities. Every year, hundreds of such institutions are closed as they are not able to meet the criteria set by governing bodies resulting in wastage of students' money.

2. **Static Curriculum for a Dynamic Environment:** Change is the only constant. The business environment is changing rapidly due to several factors such as legal, economic, and social. Still, the power of disrupting the way business function lies with Technological advances. Today, no business area is left untouched by

it. But the idea of change hasn't been able to revolutionize the curriculum of B-schools. They are still fixated on teaching age-old theories without relating it to the modern business contexts. There exists a huge gap between academics and industry, most of the time academics is not able to cater to those needs. Often times, companies have provided feedback about what they are looking for in terms of skills – communication skills, technological competency, leadership, and problem solving. The businesses need such managers who are not only able to plan long-term but also ensure the adaptability of the plans to the changing environment. For this, the institutes need to design a curriculum that caters specifically to Indian business environment. Qualities that the employers desired stressed three things for curricular change: practice orientation, teamwork, and perspective building. Further, they present guidelines that are important from the employers' perspective for building future curricular structure (Rao et al., 2014).

3. **Practical Teaching Pedagogies:** Management is all about understanding concepts in relation to business environment. No textbook is really sufficient to learn such a dynamic course as it is all about exploring. The role of teaching here should be inducing thinking among the students. The teaching should be made more of case based rather than just focusing on theories. Often, it has been seen that students know the theories but are unable to find its application in modern world. This creates a disposition among students regarding the functioning and planning of scenarios. They expect the already established theories to work without taking change into account. Moreover, the teaching should prompt students to explore more and discuss it in classrooms. This methodology develops not only understanding of concepts but also helps in visualizing these concepts in real world.

4. **Emphasis on CGPA Rather Than Skills:** The institutions are more pressed on CGPA rather than creating a balance between CGPA and skills. Most of the institutions don't conduct dedicated sessions for up skilling students and make them placement ready. Due to this, sometimes even the ones with higher CGPA are not able to fetch the desired result. Since management is more about real world, the institutes should also focus on skill building among

the students. Companies are looking for graduates who are not only good academically but skilled as well. In the race of getting higher CGPA, the task of skill development often takes a backseat which results in low employability.

4.5 REMEDIES

Employability issue among management student continues to be a major challenge in today's world. The gap between the skill set of existing graduates and those required by the corporate is nothing but increasing. The employability statistics present a grim situation and a reality often overlooked. To mitigate this crisis, changes need to be made at grass-root level. Even though changes are being made, the results are still disappointing. Nevertheless, some of the actions that can be taken are as follows:

- Since the number of institutions is increasing every year, the regulatory bodies such as UGC (University Grant Commission) should enforce strict quality control parameters. The university statistics should be analyzed in-depth to understand the hidden patterns. For instance, the placements of the institutions should be judged. Some parameters could be the jobs offered in various disciplines, location, package, the nature of job offers, etc. The competency of faculty should be evaluated on parameters other than academic qualifications like past record, research papers and academic contributions, etc.
- Regular soft skill workshops should be organized that focus on group discussions and personal interviews. The institutions should understand the need of the hour in terms of communication skills, technological skills, analytical skills, and problem solving, etc. Apart from this, students should be made aware about latest technologies and how to use them. They should be made aware about using technology as a tool for development. Students should focus on skill development rather than just acquiring hard skills as it is the combination of that will make them employable.
- Regular career counseling sessions and fairs should be held at colleges that make student aware about the job prospects. Every industry has ample of opportunities but the inability of a medium to conveyance presents a hurdle. Though internet is a good place for sharing information, it cannot create the effectiveness of

in-person meetings. Such fairs present a great opportunity for employers and students to interact with each other and understand needs and requirements of both the parties. It also helps in creating networks.

- For pursuing management degree, having work experience should be made mandatory like in the west. Approximately one-third of the management aspirants are fresher. It is seen that students with work experience exhibit better understanding of business environment and its needs. Students with work experience are able to relate business environment and curriculum better as compared to students who don't have one. This can be used as a method of quality control among students in institutions.
- Curriculum should be revised and tailored in a way to narrow the gap between academics and industry. It should be thought provoking in nature rather than consisting of just age-old theories. Furthermore, the faculty should create such an environment in class that enables students to have an open discussion among each other, taking real life case scenarios into consideration. This helps in better understanding of the subject matter and establishes relation between theory and reality.
- The teaching methodology should be revised to come up with a more practical oriented one. It should try to expand more on real time scenarios rather than theory alone. It should provide a room for thought and open discussions in classrooms. Such approach creates a better understanding of the subject matter and enhances communication skills as well.

4.6 CONCLUSION

Our country leads in having the highest number of youth population. Though our country also boasts of having highest manpower, having 'manpower' and 'skilled manpower' are two different things. Every year out of lakhs of intakes, only a handful are able to land a good job. Most of them get no return on investments. But why is that so? Well, as it turns out, it's the skills that you acquire during your graduation that creates the difference. It is said "it takes two to tango." Skill development requires investment of time and energy both from the institution as well as students. Skill development programs are only beneficial if students work to get benefit of it. On academic level, a curriculum revamp is needed that combines both hard and soft skills and also bridging academic-industry gap. The institutions can begin by enrolling students in personality development workshops and soft skill programs. Regular mock sessions consisting of group discussion, just-a-minute session, personal interview, etc., should be emphasized on so as to enhance communication skills. Companies, nowadays, want students to not only excel academically but volunteer for a social cause as well. Through these programs, companies try to evaluate your team spirit. Students should be motivated for volunteer work as it develops a sense of teamwork in them. The classroom sessions should be interactive rather than one-sided. The subject matter should be discussed in context to the current environment so as to develop better understanding about it. Such discussions also serve as ice-breaking session and enhances learning experience.

These days, the ability to reason and make sound decisions have become the crucial skills that the recruiters are looking for apart from communication skills and technical skills. Students should be familiarized with real working environment through internships, industrial visits so that they understand the requirements of corporate world. Moreover, the internship period should be around six months so as to give students ample amount of time to explore industries, businesses, and other units. This helps them to analyze and work on skills required accordingly. Also, the students should be motivated to take up projects so as to explore more about a subject matter. Moreover, they should be motivated to value skills and employability and adequate training should be provided so as to convert it to employment.

Employability issues cause one major problem which is underutilization of manpower. This results in wastage of time and resources of students and thus yielding no valuable results. Even the students should start understanding that the purpose of this degree is to explore. It's time we acknowledge that we have been doing it all wrong. We need to shift focus from employment to employability. We need to develop skills rather than just the CGPA. Changes are needed – starting from curriculum to pedagogy of teaching. Overall personality development should be the target. Thus, tackling the employability issues requires efforts from all the stakeholders, i.e., the students, institutions, and governing bodies. Only then we will be able to make our manpower a skilled manpower.

KEYWORDS

- **career counseling sessions**
- **employability**
- **employability issues**
- **employability skills**
- **interdisciplinary management education**
- **skilled manpower**
- **university grant commission**

REFERENCES

Berntson, E., (2008). *Employability Perceptions: Nature, Determinants, and Implications for Health and Well-Being.* Doctoral dissertation, Psykologiska institutionen.

GP, S. (2021). Employability of Management Graduates from Tier 2 and 3 Institutions. *AIMS Journal of Management.*

Goleman, D., (2011). The brain and emotional intelligence: New insights. *Regional Business,* 94–95.

Harvey, L., (2000). New realities: The relationship between higher education and employment. *Tertiary Education & Management, 6*(1), 3–17.

Helen, K. *What is Employability and Why is it Important.* https://www.futurelearn.com/info/courses/prepare-career-success/0/steps/21285 (accessed on 20 February 2024).

Hillage, J., & Pollard, E., (1998). *Employability: Developing a Framework for Policy Analysis.*

India Skill Report, (2021). *Less Than Half of Indian Graduates are Unemployable, BTechs Have Highest Skills.* https://indianexpress.com/article/education/less-than-half-of-indian-graduates-are-employable-btechs-have-highest-skills-report-7194119/ (accessed on 20 February 2024).

Jetter, M., Satzger, G., & Neus, A., (2009). Technological innovation and its impact on business model, organization, and corporate culture–IBM's transformation into a globally integrated, service-oriented enterprise. *Business & Information Systems Engineering, 1*(1), 37–45.

Nawaz, N., & Reddy, K., (2013). Role of employability skills in management education: A review. *ZENITH International Journal of Business Economics & Management Research, 3*(8). Available at SSRN: https://ssrn.com/abstract=3520312 (accessed on 20 February 2024).

Pallavi, P., & Ranbir, S., (2020). A study on perception of employer brand value in Chandigarh City. *International Journal of Advanced Science and Technology, 29*(11s),

1061–1066. Retrieved from: http://sersc.org/journals/index.php/IJAST/article/view/20771 (accessed on 20 February 2024).

People Management Report, (2019). *Studies Reveal the Subtle Ways Managers Sabotage Their Teams.* https://www.predictivesuccess.com/wpcontent/uploads/2020/03/2019_People_Management_Report.pdf (accessed on 20 February 2024).

Ramaswamy, N., (2016). Partner, education and skill development sector, KPMG, 2016. *25 Years of Reforms: What 1991 Reforms Have Taught B-schools.* https://economictimes.indiatimes.com/industry/services/education/25-years-of-reforms-what-1991-reforms-have-taught-b-schools/articleshow/53308482.cms?from=mdr (accessed on 20 February 2024).

Rao, T. V., Saxena, S. S., Sherry, C. V., Narendran, R., Bharathan, K., & Jajoo, B. H., (2014). *What Employers Value in the MBAs They Recruit: Rebalancing the Management Education Curriculum.*

Ruby, S., & Pallavi, P., (2020). A study on service quality of banks in tricity. *International Journal of Advanced Science and Technology, 29*(11s), 1067–1077. Retrieved from http://sersc.org/journals/index.php/IJAST/article/view/20773 (accessed on 20 February 2024).

Salovey, P., & Mayer, J. D., (1990). Emotional intelligence. *Imagination, Cognition and Personality, 9*(3), 185–211.

Saunders, V., & Zuzel, K., (2010). Evaluating employability skills: Employer and student perceptions. *Bioscience Education, 15*(1), 1–15.

Van, D. H. C. M., & Van, D. H. B. I., (2005). The development and psychometric evaluation of a multi-dimensional measurement instrument of employability—And the impact of aging. In: *International Congress Series* (Vol. 1280, pp. 142–147). Elsevier.

Vanitha, P., & Jaganathan, (2018). A study on enhancing employability skills of graduates in India. *International Journal of Trend in Scientific Research and Development, 2,* 947–951. 10.31142/ijtsrd9561.

Yorke, M., (2004). Employability in the undergraduate curriculum: Some student perspectives. *European Journal of Education, 39*(4), 409–427.

CHAPTER 5

Interdisciplinary Learning: Probing into the Role of Digital Technologies in Management Education

ARUSHI SETHI,[1] PUNEET KUMAR SETHI,[1] and AHMET ARIF EREN[2]

[1]*Assistant Professor, Chandigarh Group of Colleges, Landran, Punjab, India*

[2]*Professor, Nigde Omer Halisdemir University, Department of FEAS, Nigde, Turkey*

ABSTRACT

The traditional education system is also known as customary education in which the main motive of the education system is to pass on the information to the next generation. The system basically includes the knowledge about the customs, traditions, and religions. Prior to the digital era, the learners have to get themselves enroll in the schools or either library. But the main drawback of the traditional education system is that the learners are not able to access the current information which is happening around the world. With the introduction of modern education system, the people are able to access the current information and society is moving towards a knowledge-based society. Advancement in technology has played a greater role in the new education system because it has opened many areas of the learning. With the help of digital technology, the presence of the students in the classroom can be increased and it also help the teachers to improve their lesson plans which facilitate personalized learning. The traditional education system is a teacher-based center system in which the information

Interdisciplinary Approaches in Management Education.
Tanusree Chakraborty, Ramandeep Saini, Rajiv Khosla, Nandita Mishra, and Ashok Natarajan (Eds.)

is acquired within the four walls of the classroom, and it has a number of the advantages like social exposure, extra-circular activities and proper guidance can be taken from the teacher. It comprises of the conventional education system in which they use the tools like blackboards, chalks, pens, and papers. Students acquire information from the teachers present in the classroom and get direct assistance from them. The content prepared by the teachers for delivering the lecture is prepared either with the help of textbooks or the self-notes prepared by the teacher with the help of textbooks. The student has to get himself enrolled in the schools to acquire the education. A teacher should have a significant amount of knowledge to educate students. It is necessary for the students to be regular in the classes and have to retain the information which has been delivered by the teacher. Apart from the formal education the student can also acquire informal education through libraries, newspapers, and books. The teaching is considered as a respectable profession as it teaches every citizen of the country. With the emergence of industries, the demand for skilled people has been increased and the traditional education system was not able to cope up with the demand and thus the need arises of a new and modern education system.

5.1 INTRODUCTION

The rapid growth of digital system has revolutionized the people thinking, their behavior, their communication, and their lifestyle too. This revolution has created the new knowledge in the people and also created awareness among them. It has restructured the way people conduct business and economic practices. It helps in providing speedy delivery to healthcare centers, and a new vision for environmental sector. It has also helped in improving the standard of living of many people around the world. It has also helped the people to improve communication between people and also help to resolve the conflicts among them and to attain world peace.

The digital revolution has been referred as the advancement of technology from mechanical devices to the digital technology available today. This started in the year 1980's and it is ongoing. The digital revolution has been marked as the beginning of the new era. The development of technology started with one basic idea, i.e., the introduction of Internet. The brief outline of the digital revolution has been mentioned below:

- 1947–1979: In 1947, the transistor was introduced which lead to the way of advanced digital computers development. The military, the government organizations use this computer during the early 1950s and 1960s. This eventually would lead to the creation of the World Wide Web.
- 1980s: In this era the computers have become familiar and by the end of the decade the computers have become a necessity for many jobs. The first cell phone was also introduced in this era.
- 1990s: In the 1990s the Internet has become a part of everyone's life. By 1992, the World Wide Web has also been introduced, and by the end of 1996 the Internet has been commonly used in business day to day operations.
- 2000s: In this decade, the revolution of digital technology has begun to spread all over the world. The mobile phones were also introduced, and it can be commonly seen in the hands of people. The Internet usage has also been growing rapidly and the number of users using Internet has also been growing. The televisions have also been upgraded because now they are able to connect through digital signals.
- 2010: In this decade Internet has been used by the 30% of the world's population. The mobile phones have also been commonly used by 70% of the population. In this the tablets have also been introduced and the number of applications has been introduced which are commonly used by everyone. The people are getting more aware about the benefits of computers and most of their work is done with the help of computers.
- 2019: Undoubtedly, 2019 had brought progress, prosperity, and upgradation to Robotic automation. Autonomous things like robots, drones, and autonomous vehicles have been developed along with AI Solution.
- 2020–2021: The era of 21st century is also known as the era of technology. Without technology we cannot imagine our life. In today's world technology plays a greater role. It has been seen as the basis of an economy. The technology has made our work easier and smoother. The impact of technology can be seen every field and one such field is Education.

5.1.1 HIGHLIGHTS

- Advancement in the pedagogical sector has greatly affected the education sector.
- Technology has greatly helped the higher education (HE) to overcome the limits of space and time.
- Digital technology has introduced various educational resources to teach the students like educational videos, course materials, etc., which are generating a large amount of usage data.
- Digital technology has replaced the traditional paper-based books. Many printed books are now available in the form of e-book.

5.2 TRADITIONAL EDUCATION

The traditional education system, centered on teachers imparting knowledge in classrooms, offers social exposure, extracurricular activities, and direct teacher guidance, using tools like blackboards and textbooks. In contrast, the modern education system, leveraging technology, emphasizes personalized, flexible learning through online courses and virtual classrooms, fostering critical thinking, problem-solving, and collaboration skills. It also promotes lifelong learning, preparing students for success in an ever-evolving world. Classroom teaching provides a structured environment for learning, facilitating direct interaction between teachers and students. This approach allows for immediate clarification of concepts, personalized feedback, and social interaction among peers, enhancing the learning experience. Additionally, classroom teaching promotes discipline, time management, and collaboration skills, preparing students for real-world challenges.

Traditional education as stated is the teacher-based system. This generation was dependent on their teachers for guidance and help and their awareness and skills were restricted to the knowledge shared by the teachers. Climatic changes and other personal barriers affected the student's education when students were not able to reach their institutes. Students used to travel miles to reach educational institute to gain knowledge.

Today when the world has changed and has shifted rapidly to the virtual world instead of physical world in all fields of life, there exists the need to take management education (ME) to the virtual world with the help of Internet and latest technological tools (Chakraborty & Gupta, 2019; Chakraborty et al., 2021). Today the institutes have started using many online tools to interact with their students and have started opted digitalized courses and strategies for better delivery of study material. COVID pandemic has also emerged as a blessing in disguise for the digitalization in management and other educational courses. Here when the world stopped, education didn't. Institutes have already accepted the need for great revolution due to this emergency and have totally changed the classroom learning structure into to the virtual learning. Students have also not limited their knowledge and skills to the classroom learning. They have started gaining knowledge about practical exposures which they get by use of virtual management tools and games which help in better understanding and implementation of concept.

5.3 DIGITAL ERA

Electronic era started 100 years back when the wireless electronic communication was commonly used. The information was transmitted either through telegraph or with the help of radio and the messages were passed through air, which are invisible, on radio waves. From their onwards technology has been showing a greater change. Its journey has been remarkable shifting from radio, to recordings, to movies, to televisions, to computers, to CDS, CD ROMs and the Internet.

The Internet was very effective in conveying the information and making the people aware about the international and national news. This has become an effective mode of communication. We can see a lot of changes in the education system with the introduction of technology. To add values in the teaching learning process the universities are creating suitable learning among students.

With the help of technology, the students are able to express their ideas, knowledge, and thoughts with each other through social media. On the other side, online classes have also been offered by the schools which are attracting lot of attention. Digital technology is growing the world is also becoming flattered. They are giving us the new methods of learning. This has changed the overall education system. Digital technology has not only improved the education system, but it also has introduced the new methods of teaching which are very helpful for the teachers.

5.4 POSITIVE EFFECTS OF DIGITAL MEDIA ON EDUCATION SYSTEM

The digital media has been an important part of student's social life. It has been considered as a platform for learning which has improved the capabilities of the students. In today's global era the social media has become a major influencing factor on our education system. Social media is considered as the strongest means of communication which has helped the people to connect from one place to another place. The online platform consists of Twitter, Facebook, YouTube, Instagram, and WhatsApp among many other. Social media has shown many positive impacts on our education system which include timely information, better communication, enhancing skills, learning. With the help of social media platform in the school, the students get motivated and encouraged to learn. The educational videos on You Tube, online notes, e-books, and learning through via video calling are considered as some of the major aspects which contribute to the development of education system. The education through distance learning is one of the best ways for the students to get knowledge and to learn from the recognized universities. Social media also help to enhance academic performance and also help in enhancing creative skills among students. The platform which has been provided by social media helps in improving one writing skills, social skills in achieving one's educational goal.

5.5 NEGATIVE INFLUENCES OF DIGITAL MEDIA

Digital media has also shown a negative impact on students. The students are getting addicted to the social media which causes distraction and hindrance of mind. They are becoming habitual of social media. With the extensive usage of social media, the number of students these days are more dependent on such type of platform to gain knowledge and information rather than looking out for the journals, notes, or the books. It is easy to gather information through reading habits, online and research capabilities has been reduced among the students. Excessive use of social media has also effect the physical as well as mental health of the students. Nowadays students don't take their proper meal and do not take proper rest by continuously using mobile phones, laptops, and computers. It is advisable to the parents to keep an eye on their children so that they don't get exposed to eye strain, bad posture, mental stress, and physical stress.

5.6 COMPARING POSITIVE AND NEGATIVE EFFECTS OF THE DIGITAL TECHNOLOGY

There is an old saying that if a technology has a positive effect than it has some negative effects too. From the positive point of view, the technology has modernized the education system. It has provided the smarter way of doing the education. The learning has become more interesting and intellectual. The students are learning the education very easily and quickly. Like today in the schools and colleges the smart classes have been introduced for the students and with the help of visual aids the learning has become easy. The content of delivering the lecture has also been improved by the teachers. The teachers are using different techniques and methods for teaching the students. In today's life we can see that the students are learning about our religion through television like Mahabharata, Ramayana, and many more this kind of series has been played on the television so that they can create awareness among the students. They are also helpful in enhancing our knowledge through this kind of shows. Nowadays, if the person has to travel from one place to another place than they can take the help of GPS through which they can reach to their destination easily and quickly as compared to older days they don't have to take help from the people to reach to their destination. The technology has grown so faster that by sitting at home we can place order of anything, and it will reach our home easily whether it is related to any food item, any product or anything else. The OLA, Uber cabs has also been introduced as mobile applications with the help of which we can go anywhere around the city or within the city. Introduction of online classes among the students in the period of pandemic has shown the greatest importance of technology in our life.

If technology has some positive effects than it has some negative effects also. Social media and mobile phones may lead to physical and psychological issues, such as eyestrain, mental pressure, depression, etc. The excessive use of technology has a significant impact on teenager and developing children. Nowadays you can see that the kids who are generally at the age group of 1–3 years know how to use mobile phones and computers which are adversely affecting their minds. The overuse of technology may lead to isolation. A survey conducted in the year 2017 in which we have seen that the young adults aged 19–32 years were using social media more frequently and they found isolated at a single place. The technology has also affected our sleeping habits. Many of us sleep late at night because we are busy in texting our friends or scrolling through social media. People don't even put mobile phones away from themselves because they are so addicted to it. All of this leads to interrupted rest and overall poor sleeping habits. The late-night use of mobile phones may lead to the sleep hormone which makes it even harder to take a good night sleep.

Technology has made our lifestyle dull. Kids who play the video games constantly they will get a less time to do physical exercise. The more we spend time on hanging out on Instagram and You Tube the less time we will get time to spend the time outside and to play outdoor games.

The drawback of using social media is that it has been a constant source of distraction. If we are doing some important work that the continuously messages on our cell phones will distract our mind and our concentration too. The technology may also lead to back pain and bad posture. It is difficult for us to sit straight if we are using the cell phones. Many of us are suffering from back and neck pain which may lead to severe issues in the health. The technology has also reduced the imagination of the students. The students are not using their creative ideas, thinking and imagination rather they are more dependent on technology. In the older days we use to play on cardboards, but that day has gone. The imagination can be no longer seen in the students. For any problem they look into their tablets or mobile phones, or they jump into video games where they are provided with non-stop fun. By looking at this scenario we are worried about how our children grow older.

At the end we conclude that the digital technology has been changing our lives both in a negative and positive ways. To take the advantage of positive ways we have to overcome from the negative factors so that the technology can be used in a best possible way and also for the benefits of the students too.

5.7 USE OF SOCIAL MEDIA IN EDUCATION

The use of social media has provided the teachers and the students with the wider platform for enhancing the knowledge. In today's era the use of digital media has been increased massively. Digital media has given the various opportunities to the students and the teachers to develop new knowledge method. Through digital media we can achieve useful information and it has also become an integral part of our life.

Utilization of digital media can also be seen in the smaller cities. These cities have been enlightened with the new convention of the education system. The education system in whole has not only been able to transform the education system but also changed the thinking of the students towards their learning but also has been helping the teachers to improve their content and delivery in a bigger way.

Educators believe that progress and development of social media has played a greater role in modernizing the education system. With the introduction of smart classes, use of digital media, Internet based learning has made the learning more enjoyable and has also improved the communication between the teachers and the students.

Digital media has improved the assistance of teaching-learning process in schools and universities. The literacy rate has been improved. The educators need to discover more idea for implementing excellent practice of digital media.

5.8 FACTORS AFFECTING TECHNOLOGY IN EDUCATION

There are number of challenges that a teacher has to face in online teaching. The teacher should be well-versed with the knowledge of computer. The teacher should know how to use these technologies in a teaching. The introduction of these technologies has forced the organizations to provide training to the teachers. Sometimes teacher do not have a positive attitude towards online teaching and negative attitude towards teaching leads to great failure in the education system.

The most common barriers to technology are:

- Lack of knowledge;
- Lack of time;
- Lack of experience;
- Lack of support;
- Lack of resources.

Another barrier which has greatly affected its reliability includes hardware failures, poor internet connections, lack of updated software. Sometimes the students are living in a rural area and there is a problem of connectivity over there and he is not able to join the class and attend the classes regularly. The teacher may not have updated software through which she can take the classes. There are numerous factors which has to be faced in the online teaching.

5.9 DIFFERENT TYPES OF APPS USED IN EDUCATION SECTOR

The digital technology has introduced different types of apps in education sector like zoom, Google classroom, iCan, Kids Academy, Byju's learning, Vico Smartboard, etc., by using these apps the teaching can be made more productive and meaningful by streamlining the projects and assignments. The teachers can create classes, distribute the notes and assignments, can send the feedback, and can see number of things at one place only. By using these apps, we can also conduct conference online with number of people at one place.

In schools and in colleges these apps have played a greater role in transforming the education sector and shaping the career of the students. These apps have provided learning techniques in a modern way like by giving them challenging tasks, educational games, and puzzles. The learning through app is mostly liked by the students because these are accessible everywhere anytime. These also help the teachers to track the student performance and to give an accurate report to the students.

With the help of these apps the students can save their lot of time and they do not have to travel anywhere to attend their class. They can join the class from anywhere with the help of the link provided by their teachers. It is also the cost-effective and there is a wide range of payment options available with the students which allow them to make payment in installments or later on.

Expert lectures have also been conducted by the schools and colleges from different universities which helped the students to increase their knowledge. International conferences have also been conducted online to enhance the skills of the students. Different types of sessions have also been organized by the colleges like yoga day, Independence Day, and many more to engage the students in various activities so that they can get the education out of the box also.

5.10 DIGITAL MEDIA IN THE 21ST CENTURY

The digital media has redefined the education system in number of ways. The teachers and the students were able to participate in the classroom learning and it also has improved the communication skills. It is becoming a powerful tool of communicating, connecting, learning, and creating among students.

Digital media has transformed the education system atop the last few years. The modern methods of teaching outside and inside the classroom have been brought by the digital media. The technology has brought dramatic change in our lives. Nowadays we cannot imagine life without internet and smartphones.

Educational entities have been shifted from traditional education to modern education. New methods have been involved in the teaching methodology with the help of digital media. Computer has become a necessary base for learning. There is no need of examinations which are based on pen and paper. The learning is also no longer required to be revolved around the ability to just write, read, and cram syllabus.

The technology has become a driven force for the new change in the educational sector. The introduction of artificial intelligence (AI), robotics, etc., have revolutionized the education sector. These are the growth drivers which are changing the dynamics of employment as new understanding and skills are required to meet the future demand. The institutes are forced to introduce digital technology in their curriculum. To deal with modern businesses general skills are required to solve their day-to-day problems.

5.11 ROLE OF DIGITAL TECHNOLOGY IN COVID-19 PANDEMIC

Digital technology has supported the different sectors of the society in pandemic times in a number of ways but the sector which has been greatly supported by the digital technology in pandemic is educational sector. This sector has been a more challenging for the teachers as well for the students because now they don't have to attend the classes in a physical mode, but they have to attend the classes and have to acquire knowledge through online teaching. The students have shown a positive response through online teaching.

Data of the students has been collected by the schools and colleges through online mode. The presence of the students can be seen more in the classes. Digital technology has shown a greater path to the students to learn and to explore. During COVID-19 different projects has been assigned to the students and they have to work on that and have to present their project through online mode. Even the parents have an opportunity to participate in these kinds of activities which are not available before pandemic. They also become the part of these fun learning activities.

5.12 CONCLUSION

- Understanding the driving factors and the performance impact of digitalization on education sector.
- Practically, this study contributes towards educational sector highlighting the crucial need of digital technology.
- This chapter aims at filling the gap between the regions and demographics that have access to modern information and communications technology and those that don't have access.

In the nutshell we conclude that transformational change due to digital media has taken the education sector to a higher level. With the help of digital media every citizen of the country is able to get best qualification from the recognized universities. Technology has shown a positive impact on the education as well as the negative impact also. Students and the teachers should take this challenge positively and try to eliminate those drawbacks which are pulling them back in achieving their goals. It is therefore suggested to the schools, colleges, and the universities to introduce more and more technology in their institutes so that maximum benefit of technology can be taken out. And the slogan used "Padhega India Badhega India" is only can be achieved through digitization. The integration of technology into education has profoundly transformed the learning landscape. Students now have the ability to connect, share ideas, and collaborate with peers worldwide through various social media platforms. Additionally, schools are increasingly offering online tutorials, which have gained immense popularity for their accessibility and convenience. This digital revolution has not only changed the way we learn but has also flattened the world, providing innovative methods of education delivery. Digital technology has brought about significant improvements in the education system, introducing new teaching methodologies that benefit both students and teachers. Social media, recognized as a powerful communication tool, has facilitated connections across geographical boundaries, fostering timely information exchange and enhancing communication skills. Platforms like Twitter, Facebook, YouTube, Instagram, and

WhatsApp have played a pivotal role in reshaping education, enabling students to access educational videos, online notes, e-books, and participate in learning through video calls. The advancement in technology has expanded the horizons of learning, offering a more inclusive and personalized approach. It has increased student engagement and participation in classrooms, allowing teachers to tailor their lesson plans to cater to individual learning needs. Moreover, technology has made distance learning a viable option, enabling students to access education from renowned institutions without physical constraints. Social media's impact on education extends beyond communication and collaboration; it has also been instrumental in enhancing academic performance and nurturing creative skills among students. In conclusion, distance learning offers students a valuable opportunity to acquire knowledge from renowned universities, while social media enhances academic performance and fosters creative skills. The platforms provided by social media play a crucial role in improving writing and social skills, aiding students in achieving their educational goals.

KEYWORDS

- **COVID-19 pandemic**
- **digital media**
- **digital technology**
- **education sector**
- **management education**
- **negative influences**
- **online teaching**
- **virtual management**

REFERENCES

Chakraborty, T., & Gupta, D., (2019). Examining dynamics of emotional intelligence and entrepreneurial intentions among management and non-management students. *The International Journal of Indian Psychology*, *7*(2).

Chakraborty, T., Tripathi, M., & Saha, S., (2021). The dynamics of employee relationships in a digitalized workplace: The role of media richness on workplace culture. In: *Critical Issues on Changing Dynamics in Employee Relations and Workforce Diversity* (pp. 175–205). IGI Global

CHAPTER 6

Teachers as Strategic Thinkers and Designers of Learning in the Context of Interdisciplinary Management Education

GEETHA RAJARAM[1] and AHMET ARIF EREN[2]

[1]*Principal and Director, Global Institute of Management Sciences, Bangalore, Karnataka, India*

[2]*Associate Professor, Nigde Omer Halisdemir University, Department of FEAS, Nigde, Turkey*

ABSTRACT

Great teachers are strategic thinkers and designers of learning who aim to create an environment where students feel empowered and enthusiastic about learning. This chapter explores the ability to comprehend and apply learned concepts, make decisions in complex situations, and engage in linear progress and vertical learning. It also emphasizes the importance of identifying teachers' core competencies and developing a competitive edge, as well as cultivating a strategic mindset to anticipate and prepare for different outcomes when implementing lesson plans. The methodology is based on the real-time experiences of various teachers in higher education (HE). The chapter encourages teachers to envision themselves as designers of learning, utilizing a spectrum of teaching strategies with the aid of Strategic Maps. It concludes with a well-designed Self-Development Plan for teachers.

Interdisciplinary Approaches in Management Education.
Tanusree Chakraborty, Ramandeep Saini, Rajiv Khosla, Nandita Mishra, and Ashok Natarajan (Eds.)

6.1 INTRODUCTION

"Teachers are leaders, because they lead the change."
—Payal Ninjiani, Globally Acclaimed Leadership Expert, New York Award-Winning Author.

Teachers, who are often strategic thinkers, continually generate and compile exciting new ideas for their students. Developing new teaching models using technology-enabled course delivery methods turns traditional classes into engaging learning spaces.

In the recent past, universities and colleges have undergone significant changes, impacting the way teachers teach. To establish a strong presence and motivate learners, it is crucial to ensure they feel cared for. Personal efficacy and sustained curiosity play a vital role in building a strong relationship.

> "Strategic thinking is simply an intentional and rotational thought process that focuses on the analysis of critical factors and variables that influence the long-term success of a business, a team, or an individual" (Center for Management and Organization Effectiveness, 2019).

> "If you lead organizations that have adopted agile methods, you know that creating the right environment for your agile teams is crucial. This includes accelerating learning by integrating users into a fast-learning loop" (Koning, 2019).

Another researcher has adopted notions of hierarchy, such as first-order competence, which comprises customer and technological competencies; integrative competence, which is the ability to combine the previous competencies; and second-order competence, which is the ability to create first-order competencies (Danneels, 2002).

The authors have identified common features of strategic thinkers, distinguishing them from other types of leaders, such as executors or strategic planners. They found that strategic thinkers are characterized by the following traits: (i) a strategic perspective, which involves a long-term approach and the ability to plan ahead, a holistic view of the company in its environment, and openness to change leading to innovative solutions; and (ii) a reflective thinking style, which includes awareness, insight, reflection, and double-loop learning (Piórkowska et al., 2021).

In recent views of classic strategic planning processes, emphasis is placed on the need for strategy to emerge rather than be solely planned, the involvement of various stakeholder groups, and the importance of executing strategic initiatives (Liedtka & Kaplan, 2019).

> "Combined with increased complexity and multi-layered requirements of modern markets, there is a demand to accelerate strategic planning cycles and design them more dynamically and agile" (Lund Storm et al., 2018).

> "One of the central competencies in successful strategic management is strategic thinking. Whereas strategic planning is formal and analytical, strategic thinking is more creative, intuitive, and reflective" (Pillai et al., 2020).

The designers' mind, which understands framing and reframing the problem space, sees the context as being dynamic, and takes a holistic view, integrating a systems perspective to shape the future (Buehring & Bishop, 2020; Pillai et al., 2020).

Like any other leader, teachers as strategic thinkers require research, analytical thinking, problem-solving abilities, communication skills, leadership, and decision-making.

Effective educators are strategic thinkers and learning designers. They strive to cultivate an empowering and engaging learning environment for students. For a long time, the role of teachers in designing the curriculum, innovating, and designing learning was limited. Now is the opportunity for teachers to design a learning space for their students within the *learning framework.*

The learning framework essentially relies on three things: enabling students' self-discovery, building relationships, and self-discovery through co-inquiry (Figure 6.1).

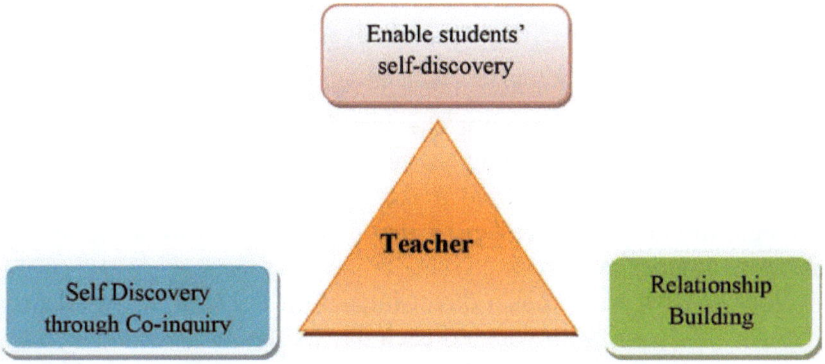

FIGURE 6.1 Learning framework.

Source: Developed and designed by the author.

6.2 POSITION OF LEVERAGE

The learning framework introduces us to three levers: enabling self-discovery among students, relationship building, and self-discovery through co-inquiry. Your skills, teaching methods, and capabilities can be leveraged to create value. It is the leverage versus the unleveraged factors that a teacher can outperform in a class. Teachers need to find a position of leverage. Teaching is a preference you make and a proficiency you develop. What you want is to organize your class for effective learning. Therefore, always factor your energy, preparedness, and attitude into every class (Figure 6.2).

FIGURE 6.2 Your class = f{your energy, attitude, and preparedness}.

Source: Developed and designed by the author.

6.3 STRATEGIC THINKING: LINEAR AND VERTICAL LEARNING PROCESS

Your teaching methods are not one-size-fits-all strategic plans in search of excellence. By integrating strategic thinking into your work, you would be more organized and make use of the possible opportunities. Strategic thinking requires agility and determination in choosing a plan and executing it.

Your state of being has a direct impact on the overall class learning. The aura you carry with yourself each day to class has an impact on your classmates. While supporting learners is instrumental, you should use an integrated approach where the learner can evolve. It is important to look for linear and vertical progress of a learner, which is also sustainable. By using linear and vertical learning methods, you will realize that learners can actually develop competencies such as communication, enabling them to respond effectively and develop technical expertise. Essentially, horizontal learning enables the development of knowledge, skills, and behavior. Linear progress of a learner could be the ability to understand a concept in a given area or situation. Vertical learning is about the transformation of how the learner thinks, applies the concepts learned, and makes decisions in complex situations. Great teachers should focus on action, whether it is the action of learning basic concepts or the action of identifying patterns in data and making decisions (Figure 6.3).

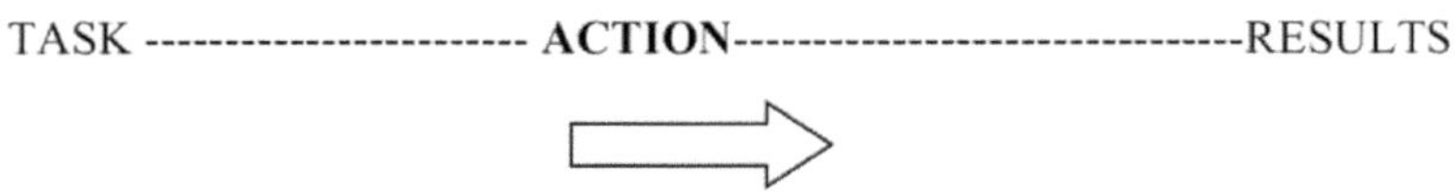

FIGURE 6.3 Task–action–results.

Source: Developed and designed by the author.

6.4 CORE COMPETENCIES TO REMAIN AHEAD OF THE CURVE

Teachers should periodically examine their strategic initiatives to ensure effective execution, review, and sustain their efforts. In order to stay *ahead of the curve,* it is necessary to identify core competencies and develop a competitive edge. Competency development is crucial for the sustainable progress of any teacher. The competencies of a teacher are broadly divided into:

1. **Pedagogical Competencies:** These include knowledge, technical skills, effective communication, instructional delivery methods, and formative assessment.
2. **Professional Competencies:** These include classroom management, tailoring resources to suit the needs of your class, setting clear expectations for the class, and documenting progress.
3. **Logic-based Competencies:** These are based on your reasoning, experience, and discussions with senior faculty members, which enable you to address certain issues or make decisions in certain situations.

4. **Evidence-based Competencies:** These are identified through empirically based data or information gathered from within the institution, such as a student satisfaction survey.

When looking for core competencies, you have to think of it as an association of mutually reinforcing talents, skill sets, and experience that align with your intent to plan your course delivery. Identifying and developing core competencies is crucial for teachers in adding greater value to their learners. Some of the core competencies that you could identify and develop are:

- Knowledge (subject knowledge, tools, and techniques, etc.);
- Communication skills (listening to learners/sharing information/enable questioning);
- Interpersonal skills (emotional connect/support/respect/empathy);
- Mentoring and feedback (providing constructive feedback/mentor learners);
- Strategic thinking (plan/visualize/deploy with agility/reasoning).

You can rate yourself on each of the core competencies using a three-point scale:

- Essential;
- Enhance; and
- Expert.

When each of the core competencies is ranked, this rating would help you identify your core competencies and enable you to determine which of these competencies need improvement or development. Additionally, it would help you identify the competencies in which you are an expert and allow you to further enhance them.

6.5 BUILD A STRATEGIC MINDSET

Build a strategic mindset to review each core competencies in brief:

- Key elements (what are the essentials for class delivery);
- Development needs (what you are good at and what needs improvement);
- Tools for improvements (attend workshops/FDPs, etc.);

- Make notes as points for reflection, considering feedback and revisiting your teaching methods. Do not be worried about experimenting with new skills.

Strategic mindset is the ability to anticipate and prepare for various outcomes when executing your plan in class. Instead of dwelling on 'what went wrong?' approach it with forethought and consider:

- What went right? (accelerate);
- What needs to be tweaked? (start);
- What went wrong? (stop).

When you identify 'what went right,' accelerate the idea or teaching method. When identifying 'what needs to be tweaked,' start the process of correcting the pedagogy or method of collecting teaching material, or even the assessment methods. This sounds pretty useful, doesn't it?

The best way to develop a strategic mindset is to give yourself time to think. Think from different perspectives. Review the past and envision the future, which would enable you to design your spectrum for effective teaching.

6.6 DESIGN THE SPECTRUM OF TEACHING WITH STRATEGIC MAPS

Strategic maps are effective and efficient ways to explain how you design, deliver, conduct, and drive the desired learning outcomes of your course. This helps to create a competitive advantage for you and optimize the relationship with learners.

1. **The Spectrum of Teaching:** The spectrum of teaching begins with a chain of decision-making based on three key elements: Insight (range of vision), Foresight (anticipate and plan), and Prudence (foresight). Every decision taken sets a path for the next course of action.

    - **Spectrum 1:** Estimate the Requirements of Your Class and How it is Evolving.

      It is a high-level plan that sets the platform for your class, serving the learners throughout the academic cycle. This helps you prioritize your roadmap with the right initiatives and find ways of effectively using time and resources. The first step is to set the learning objectives of the course. Your strategy needs to deliver something of value to your learners and create the best conditions for them to learn and flourish.

You would have learners who are diverse in culture, age, location, and learning abilities. There could be extrinsic and intrinsic differences among them. *Extrinsic differences* are observable and help you determine their requirements. *Intrinsic differences* are learners' characteristics or unobserved differences, such as their level of passion to learn, ability to fit with others, or tolerance.

First, understand and assess the features that learners value the most. Learners feel valued when they are appreciated or respected in learning spaces, when there is an atmosphere of inclusion, intrinsic motivation, engaging course content, and positive class dynamics (Figure 6.4).

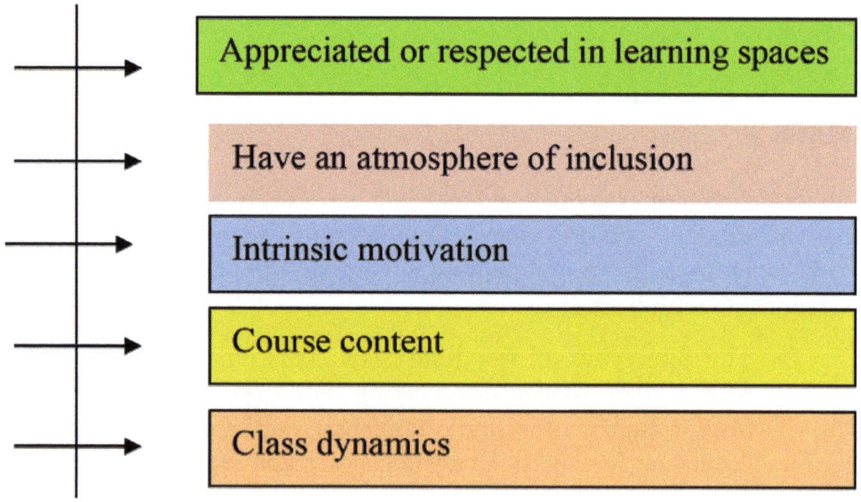

FIGURE 6.4 What learners value most?

Source: Developed and designed by the author.

It is not enough to simply estimate the class requirements based on the course content and resource needs. You must truly understand the needs of your learners, the challenges they encounter, and their capacity to learn. Pay closer attention to how the learners' needs are evolving.

Follow these steps to estimate your class requirements:

- What features/attributes do the students value the most in your teaching?
- Rank each feature by its priority.
- Monitor the useful features in comparison to the value they provide to the learners over time.

2. Five Questions to Consider When Estimating Class Requirements:

i. How will you create value in today's class?
ii. What do learners expect you to accomplish?
iii. How will you get some early momentum in the class?
iv. What is the landscape of the class? and
v. What skill do you have, or you need?

- **Spectrum 2: Create Courses with Useful and Valuable Features**

  While designing courses, as teachers, we need to experiment. This may be unsettling for both the teachers and the learners. Learners need something to believe in. Keep these questions in mind while designing a course and try to answer them:

  - How can your course help learners? (useful)
  - What features makes your course unique? (valuable)

  This brings us to another intriguing concept: the distinction between useful features and valuable features that learners seek in a class. Useful features are practical and beneficial for securing a job, while valuable features greatly enhance their learning experience and impact their job performance. Useful features are typically tangible, whereas valuable features are intangible in terms of a learner's performance.

6.7 SOME TIPS FOR EFFECTIVE DESIGNING OF YOUR COURSE

- Start with the end in mind;
- State the learning objectives;
- Emphasize on the value of the course;
- Create an accurate roadmap and update regularly;
- Learn and train yourself;
- Roll out your class;
- Enable on-boarding of learners;
- Create a strategy for scalability;
- Leverage your knowledge and skills;
- Articulate your value proposition;
- Please gather all the students in the class and establish communication plans;
- Ensure that the learners are informed about the learning objectives and outcomes;
- Let's explore possibilities to improve teaching and decide on the next steps.

6.8 DESIGN A VALUE PYRAMID

The courses designed deliver four primary elements of values that concentrate on for kinds of need: operational, relationships, responsible and impactful. More the elements provided, greater is the impact of learning (Figure 6.5).

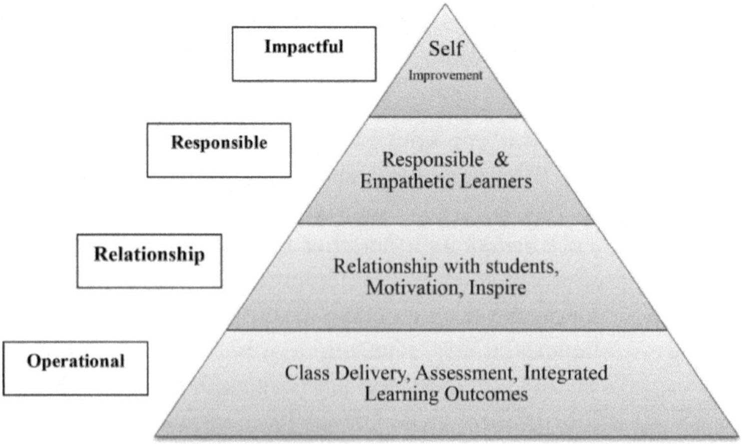

FIGURE 6.5 Value pyramid for teachers.

Source: Developed and designed by the author.

The elements of value in the value pyramid describe the power of these elements. Learners weigh their perceived value against its usefulness. How can teachers actively manage value to deliver more, whether functional or emotional? The building blocks of value exist in creating personal values like life-changing element motivation and self-improvement.

- **Spectrum 3: Build Your Design Toolkit and Communicate Your Roadmap**

  Architecting a great class that makes learners care about your roadmap does not happen by chance. Instead, it is an intentional effort intertwined into your strategy. Use the roadmap to communicate your learning objectives in a way that shares your ideas and initiatives to create a better learning space in your classes. Your delivery style becomes the vehicle for conveying your teaching style.

  Wear your 'delivery belt' and map out your journey. It is important to design a framework through a systematic study of the teaching process and requirements to avoid failures.

  Framing the right delivery makes insights and learning meaningful. It gives learners the power to understand, connect, and align with your teaching methods. A well-crafted course helps learners to:

  - Dispel confusion;
  - Increase understanding;
  - Improves engagement;
  - Inspire them to achieve more.

6.9 DO NOT FORGET TO INCLUDE THE 'WHY' AND 'HOW' WHEN DESIGNING YOUR ROADMAP

- **Step 1: Answer the 'WHY':** Your 'WHY' sets you apart from others and helps learners choose your course. Every course is important for a learner; the idea is to make learning exciting. Answer your WHY. Why are you doing this? Why will learners take up your teaching?

  Here begins the journey of your teaching.

The 'HOW' of your class brings life to your course. It describes how learners can hope to accomplish and how you plan to do it.

- **Step 2: Simplify:** Simplifying your teaching methods is crucial for engaging learners. Avoid distracting them with unnecessary details and focus on the core of learning.
- **Step 3: Remember the 'Star' is the Learner:** The learner should be the center of attention in your class. Therefore, having a deep understanding of the learner helps in designing and structuring your class.
- **Step 4: Make it Relatable**:
  - **i. Develop Empathy:** Who is the user you are describing?
  - **ii. Establish Context:** What is the learning objective that the course aims to achieve?
- **Step 5:** Consider Your Learners: The way you classify and present your roadmap depends on your audience. Focus on purpose, learning, and action. If you want learners to remember what you said, understand their unique perspectives. You can prompt learners to feel more in control of their learning and help them set learning goals.
- **Step 6:** Aim to Engage: Your class should be gripping, emotionally engaging, and unforgettable. What would you say or teach if you had half the time to teach and at the same time make it easier for learners to learn? Enable independent thinking among the learners and increase collaboration through group activities and teamwork.
- **Step 7:** Provide a Vision: Be consistent in your teaching methods. Provide a vision, direction, clear priorities for learners to focus on, and passion to achieve.
- **Step 8:** Promote Values: Make sure you promote values, as we all believe that teachers are catalysts for change.

6.10 THE THREE IS FOR DESIGN

While designing the roadmap, keep three 'Is' in mind.

Innovate —————— Implement —————— Influence

Source: Developed and designed by the author.

As designers, teachers can symbolize innovative ideas and concepts, set the mood to facilitate discussion and learning. Teachers can influence learners, curate a learning path around the concepts, connect the content, and conduct the class.

6.11 THE TRAVEL PATH

The travel path can be designed to find solutions, leverage opportunities, address problems, and initiate positive changes that would help learners adapt and position themselves for success. On the travel path, you need to reach and pass various milestones.

- **Milestone 1: Awareness:** You need to be observant and aware of the landscape of your class, learners' reactions, and expectations from your class. This will enable you to anticipate the potential needs of your learners and maintain an outlook for continuous improvement.
- **Milestone 2: Clarity:** You have to clearly define the learning objectives and articulate the problems in the class, gaining support for action among the learners. You can help learners identify what is important and why while learning.
- **Milestone 3: Inclusivity:** Learners will be full of ideas and innovative thoughts, so be open to them. Create an atmosphere of inclusiveness, collective thoughts, and leverage them for better learning.
- **Milestone 4: Empathy:** Know how to engage, motivate, and inspire learners. Establish strong relationships with them and also among them in the class. Encourage respect for each other's ideas and be empathetic.
- **Milestone 5: Supportive:** Encourage learners to explore, experiment, and question the teacher or ideas or concepts discussed in the class.

- **Milestone 6: Agile:** Take quick decisions and reorient the teaching methods whenever required. The teaching methods must be flexible and agile to ensure that the learners benefit.

Finally, while designing an effective roadmap, incorporate some 'excitement features' that you can invest in to increase learners' delight. Include them and continue to invest in them for a dramatic learners' delight. Communicate your teaching style with these features. See the 'magic' when learners opt for only your course (Figure 6.6).

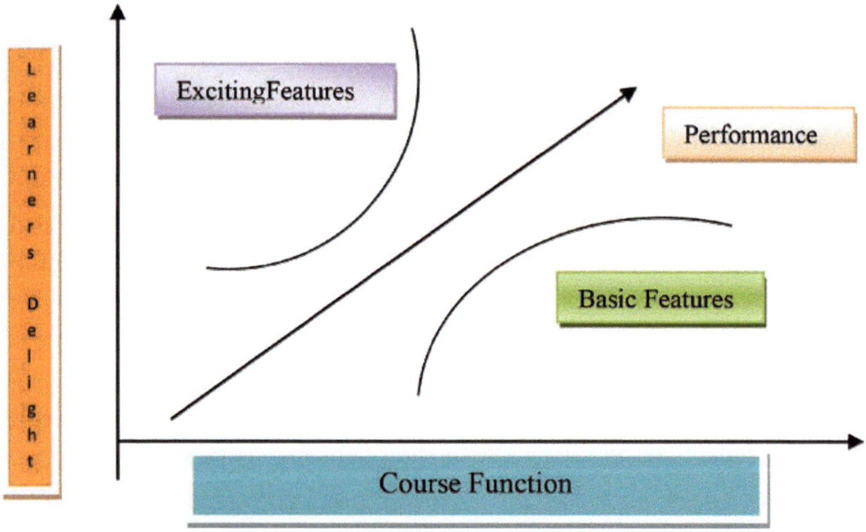

FIGURE 6.6 Features in course delivery.

Source: Developed and designed by the author.

In the meantime, it is important that learners appreciate the efforts put into designing exciting features of the course. They may be aware of the intention behind the course design or may be confused about it. They may even have an understanding of the learning objectives and how helpful the course is for them. This could be their perception. Therefore, it is important to address how to effectively measure the teachers' intention versus students' perception. At times, what the teacher intends may not be realized by the students. What might the teachers be missing? Is it due to an overload of information that students are unable to grasp the intended purpose of the course?

This leaves us with a question: Can the intention with which the teacher designs a course truly be deployed and understood by the learners? Let us take a look at the teacher intention and student perception matrix. In which quadrant do you think you fit? Does it really work? (Figure 6.7).

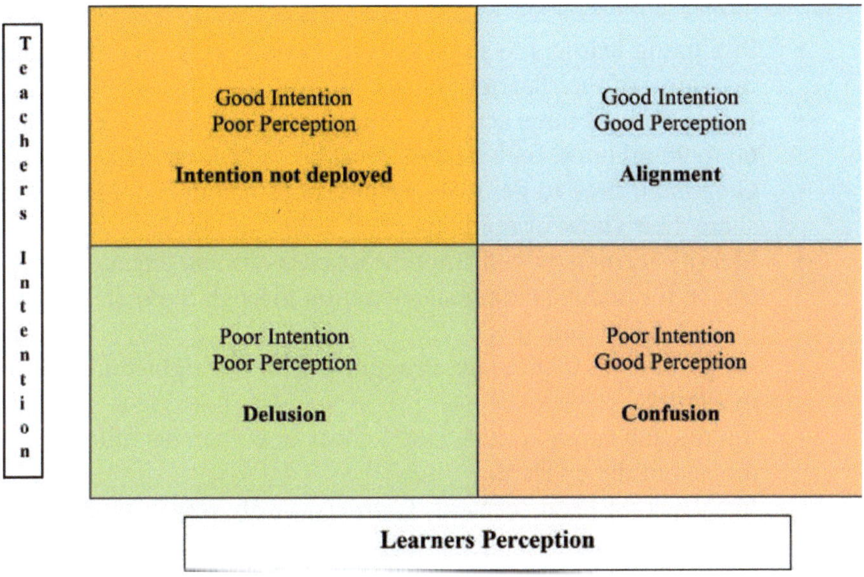

FIGURE 6.7 Intention–perception matrix.

Source: Developed and designed by the author.

When examining the intention-perception matrix, if the teachers' intention and learners' perception are poor, it can lead to delusion and uncertainty in the learning process. If the teachers' intention is good but the learners' perception is poor, the intended purpose of the course may not be realized. Additionally, if the teachers' intention is poor but the learners' perception is good, it can result in confusion during the learning process. This brings us to the fourth quadrant, where the teachers' intention aligns with the learners' perception, creating a sense of harmony and delight in the learning experience for both teachers and learners.

6.12 PLAN FOR LEARNERS ENGAGEMENT

In the recent past, universities and colleges have undergone changes, as well as the way teachers teach. To establish a strong presence and motivate learners, it is important to ensure they feel cared for. Personal efficacy and sustained curiosity play a crucial role in building a strong relationship.

- Prompting learners to set learning goals can help them feel more in control of their learning.
- Please think of three activities that you will be doing over the next 60 to 90 minutes. Include both must dos and could dos.
- Give them time to get together in groups and encourage them to share their views or opinions.
- Identify learners who disrupt the discussions and engage them.
- Check for understanding and attention through 'cold calls' (question students randomly).
- Help students if they are struggling with the right things (team dynamics).
- Submit and summarize the top 3 ideas or comments of learners at the end of the concept.
- Feedback vs Feed forward: Learners hardly revisit their assignments to check for suggestions or comments. So, share assessment criteria. Inform them what is expected and let them know the grading rubrics and mark accordingly.

6.13 ADOPT THE DST MODEL TO REVIEW YOUR TEACHING METHODS

Table below introduces the DST (Done, Stop, Tweak/Rewrite) model as a tool for educators to review their teaching methods. This model provides a systematic approach for reflecting on teaching practices, identifying what is working well (Done), areas to discontinue (Stop), and aspects that need adjustment or complete overhaul (Tweak/Rewrite). By employing the DST model, educators can enhance the effectiveness of their teaching and improve student learning outcomes.

D	Done	What have you done since the previous class?
S	Stop	Is anyone stuck?
		What resources or guidance is required to help you move forward?
T	Tweak	What modifications do we need to make in order to increase the learning impact?

Source: Developed by the author.

6.13.1 THE DILEMMA OF COMPLIANCE VERSUS STRATEGY

Like any other strategic thinker, you are faced with compliance and strategy. You may be looking at 'compliance' to complete the curriculum on time and assess the learners accordingly. At the same time, you look for opportunities to apply your teaching strategies and add value.

6.14 STOPOVER IN YOUR JOURNEY

Points of reflection help you pause and reflect on the changes required to manage the evolving requirements of learners. Positioning yourself as a strategic thinker would enhance your role as an important teacher in your organization. Teachers should periodically reflect on their initiatives to ensure effective execution, review, and sustain their efforts. This would give you the power to understand, connect, and align with the emerging needs of learners.

6.15 POINTS OF REFLECTIONS

1. Set aside time to reflect on your plan for teaching the course.
2. Prioritize tasks such as collecting reading materials or resources and creating a course plan.
3. Design your pedagogy, considering options such as class discussions, case studies, simulations, presentations, and gamification techniques.
4. Discuss with senior faculty members to explore better methods or techniques.
5. Lay the groundwork for your class.
6. Design a self-development plan.

- **Spectrum 4: Create a Self-Development Plan.**

  After estimating the requirements of your class, creating courses with useful and valuable features, building, and communicating your roadmap, it is important to have a self-development plan to evolve and progress. Develop cycles of reflection and find answers to these questions.

  - What should you keep going?
  - What should you stop doing?
  - What should you tweak for greater impact?

  What should you stop, start, and accelerate?

To start, create a RAP (rapid action plan) for self-development. Seek answers to the following questions:

1. **How Can I Create Value for My Class and Myself?**
   - What motivated you to choose this role as a teacher?
   - What are the expectations of key learners that you need to fulfill?
   - How will your progress be evaluated?

   Remember all this may not be important when you choose this role. All this evolves over time.

2. **How Should I Conduct Myself?**
   - Understand the essential norms of being a teacher.
   - Familiarize yourself with the academic field's culture.

   Behavioral Aspects

3. **Whose Support is Crucial for My Improvement and Growth?**
   - Understand the landscape for personal growth.
   - Learn how to navigate towards growth.
   - Identify whose support is crucial and why.

   Build your Network

4. **How Can I Achieve Early Wins or Build Momentum?**

    - Making quick, tangible improvements will create a sense of momentum.
    - Building credibility for your skills.
    - Accelerating your learning and making significant changes in your workspace.

    Momentum

5. **What Skills and Abilities do I Need to Develop to Excel in This Role?**

    The skills and abilities that have brought you to this point in your teaching career may become outdated and require reskilling over time. As Payal Ninjiani suggests, 25% of your career is based on your skill sets, while the remaining 75% is influenced by the person behind those skill sets – that is, you.

    - Continuously improve your skill sets.
    - Make choices that foster personal growth.

    Skill sets

We have a model for self-development:

6.16 DESIGN A SELF-DEVELOPMENT PLAN

Design a self-development plan, prioritize yourself, be a lifelong learner, make your stakeholders your storytellers, empower yourself, be a good listener and imagine where you want to see yourself in future (Figure 6.8).

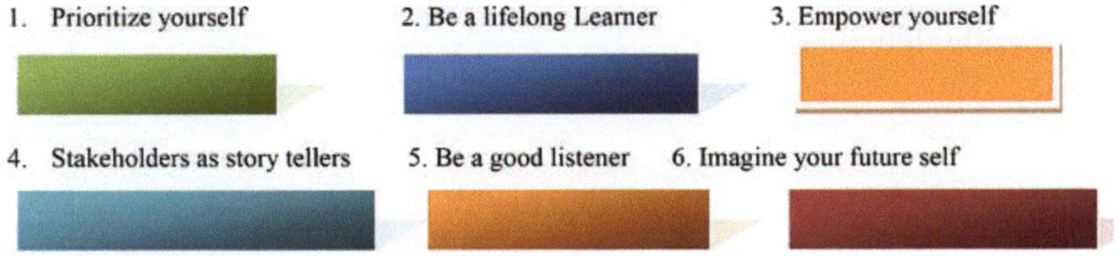

FIGURE 6.8 Designing a self-development plan.

Source: Designed and developed by the author.

Teaching is not only about having the 'right conversations;' it goes far beyond speaking and listening in a classroom setting. The key components of effective teaching involve the creation of synergies within the class, incorporating pedagogical richness, analytical rigor, and insightful analysis. It is crucial to facilitate the transferability of learning objectives through intellectual contributions and discussions by the teacher. The impactful learning experience is achieved through simulations, case studies, field work, presentations, class discussions, timely submission of reports and assignments, which ultimately shifts the mindset of the learners and promotes active learning.

6.17 KEY TAKEAWAYS

- Create an empathetic approach that puts learners at the center.
- Design classes that clearly define the learning outcomes.
- Engage students in co-creating their learning journey.
- Build a design toolkit to think through and visually display the learning process.
- Use the DST model to review the teaching pedagogy.
- Design a self-development plan.

6.18 FINAL WORDS

Learning is at its best when learners are actively engaged in the learning process. "Telling is not teaching and listening is not learning." Three things for a teacher as a strategic thinker and designer:

- Understand your audience (their pain points, their skill sets – your thoughts);
- Be a master in your field (share this with as many – your 'novel' beliefs); and
- Develop a self-development plan (your action).

You = f {your thoughts, beliefs, and action}

Teachers have the power and responsibility to create the best conditions for learners to learn and flourish. As strategic thinkers, teachers manage a heterogeneous set of students, design a roadmap, onboard learners, use the right platform, and take care of self-development.

It is important not only to teach students how to interact with technology, but also to teach them skills that won't be automated soon.

After all, a good strategy is knowing 'who you are' and 'making the right choices.'

KEYWORDS

- **core competencies**
- **learners' engagement**
- **learning**
- **self-development plan**
- **strategic thinking**
- **teaching**
- **teaching methods**
- **value pyramid**
- **vertical learning**

REFERENCES

Amason, A. C., & Schweiger, D. M., (1994). Resolving the paradox of conflict, strategic decision making, and organizational performance. *International Journal of Conflict Management.* https://doi.org/10.1108/eb022745.

Attar, M., & Abdul-Kareem, A., (2020). The role of agile leadership in organizational agility. Akkaya, B., (ed.), *Agile Business Leadership Methods for Industry 4.0* (pp. 171–191). Emerald Publishing Limited, Bingley. https://doi.org/10.1108/978-1-80043-380-920201011

Bonn, I., (2001). Developing strategic thinking as a core competency. *Management Decision, 39*(1), 63–71. https://doi.org/10.1108/EUM0000000005408.

Bonn, I., & Fisher, J., (2011). Sustainability: The missing ingredient in strategy. *Journal of Business Strategy*. https://doi.org/10.1108/02756661111100274.

Bower, J. L., & Gilbert, C. G., (2007). *Create or Destroy Your Company's Strategy.* Harvard Business Review.

Brown, T., (2019). Change by design: How design thinking transforms organizations and inspires innovation, harper Collins eBooks (1st edn.). *Business Review, 68*(3), 79–92.

Buehring, J., & Bishop, P. C. (2020). Foresight and design: New support for strategic decision making. *She Ji: The Journal of Design, Economics, and Innovation, 6*(3), 408–432.

Casey, A. J., & Goldman, E. F., (2010). Enhancing the ability to think strategically: A learning model. *Management Learning*. https://doi.org/10.1177/1350507609355497.

Center for Management and Organization Effectiveness, (2019). https://cmoe.com/glossary/strategic-thinking/ (accessed on 20 February 2024).

Edgar, W. B., & Lockwood, C., (2008). Organizational competencies: Clarifying the construct. *Journal of Business Inquiry: Research, Education, and Application, 7*(1), 21–32.

Edward, E. L. (2005). *From Human Resource Management to Organizational Effectiveness*. https://doi.org/10.1002/hrm.20059.

Erwin, D., (2002). The dynamics of product innovation an firm competences. *Strategic Management Journal, 23*(12), 1095–1121

Erwin, D., (2002). The dynamics of product innovation and firm competences. *Strategic Management Journal, 23*(12), 1095–1121. First published: 19 September 2002 https://doi.org/10.1002/smj.275.

Goldman, E. F., & Casey, A., (2010). Building a culture that encourages strategic thinking. *Journal of Leadership & Organizational Studies*. https://doi.org/10.1177/1548051810369677.

Goldman, E. F., (2012). Leadership practices that encourage strategic thinking. *Journal of Strategy and Management*. https://doi.org/10.1108/17554251211200437

Goldman, E. F., Scott, A. R., & Follman, J. M., (2015). Organizational practices to develop strategic thinking. *Journal of Strategy and Management, 8*(2), 155. https://doi.org/10.1108/JSMA-01-2015–0003.

Hamel, G., & Prahalad, C. K., (1994). *Competing for the Future* (pp. 122–128). Harvard Business Review.

Harvey, M. G., & Buckley, R. M., (1997). Managing inpatriates: Building a global core competency. *Journal of World Business, 32*, 35–57.

Interactive Teaching Styles Used in the Classroom. Available at: http://education.cuportland.edu/blog/tech-ed/5-interactive-teaching-styles-2/ (accessed on 20 February 2024).

Koning, P., (2019). *Agile Leadership Toolkit: Learning to Thrive with Self-Managing Teams*. The Professional Scrum Series.

Lewrick, M., Link, P., & Larry, L., (2018). *The Design Thinking Playbook: Mindful Digital Transformation of Teams, Products, Services, Businesses and Ecosystems*.

Liedtka, J., & Kaplan, S. (2019). How design thinking opens new frontiers for strategy development. *Strategy & Leadership, 47*(2), 3–10.

Lund, S. L. C., Willumsen, P. L., Oehmen, J., & Heck, J., (2018). Can design thinking mitigate critical strategy implementation risks? In: *DS 92: Proceedings of the Design 2018 15th International Design Conference* (pp. 1233–1244).

MINDTOOLS (Org.). *The Reframing Matrix: Generating Different Perspectives*. Available at: http://www.mindtools.com/pages/article/newCT_05.htm (accessed on 20 February 2024).

Olokundun, M., Moses, C. L., Iyiola, O., Ibidunni, S., Ogbari, M., Peter, F., & Borishade, T. (2018). The effect of nontraditional teaching methods in entrepreneurship education on students entrepreneurial interest and business startups: A data article. *Data in Brief, 19*, 16–20.

Parker, D. W., Holesgrove, M., & Pathak, R., (2015). Improving productivity with self-organized teams and agile leadership. *International Journal of Productivity and Performance Management, 64*(1), 112–128. https://doi.org/10.1108/IJPPM-10-2013-0178.

Pillai, S. V., Vipin, H., & Abhilash, V. S., (2020). Design thinking: Transforming strategic thinking towards human-centric innovations. *Review of Professional Management, 18*(2), 1–11.

Piórkowska, et al., (2021). Strategic thinkers and their characteristics: Toward a multimethod typology development. *International Journal of Management and Economics, 57*(4). doi: 10.2478/ijme-2021-0024.

Prahalad, C. K., & Hamel, G., (1990). *The Core Competence of the Corporation.* Harvard.

Rebecca, S., Nicole, E. C., Elizabeth, M. J., Diana, L. L., James, T. L., & Michael, L. R. (2018). *Review of Innovative Teaching Methods.* https://doi.org/10.1016/j.acra.2018.03.025.

Sivarajah, R. T., Curci, N. E., Johnson, E. M., Lam, D. L., Lee, J. T., & Richardson, M. L. (2019). A review of innovative teaching methods. *Academic Radiology*, *26*(1), 101–113.

Yolanda, Z. L. (2019) *Design Thinking Material.* https://www2.deloitte.com/content/dam/Deloitte/cn/Documents/about-deloitte/deloitte-cn-mmp-pre-reading-design-thinking-participant-fy19-en-181106.pdf (accessed on 20 February 2024).

CHAPTER 7

Hybrid Teaching Tools: Student Engagement Through Gamification in Management Education

PRIYA SACHDEVA[1] and M. DILEEP KUMAR[2]

[1]*Associate Professor, Amity University, Noida, Uttar Pradesh, India*

[2]*Deputy Vice-Chancellor and Professor, Faculty of Management Sciences, Nile University of Nigeria, Abuja, Nigeria*

ABSTRACT

Management education (ME) has undergone a transformative journey in this era of digitalization, embracing gamification like never before. Technology has played a crucial role in introducing gamification to ME. Gamification involves incorporating gaming elements into the teaching-learning process, creating a more immersive and enriching educational experience. It is one of the latest innovations brought about by technology. In recent times, gamification has been widely adopted in the field of ME. These educational games have made learning engaging, intensive, and insightful, taking ME to new heights. They generate interest, focus, and provide a solid learning foundation for students, learners, mentors, and researchers. The implications of gamification are global and set a benchmark for the future of ME. This chapter relies on secondary research to explore the impact of gamification on ME and students. It aims to define gamification and examine its application in education, highlighting how it has strengthened the learning experience for management students. The chapter provides valuable insights through case studies of two educational

Interdisciplinary Approaches in Management Education.
Tanusree Chakraborty, Ramandeep Saini, Rajiv Khosla, Nandita Mishra, and Ashok Natarajan (Eds.)

brands, showcasing how they have embraced gamification in ME. This study will benefit students, researchers, and academicians seeking to understand how gamification has revolutionized ME in this age of digitalization and technological advancement.

7.1 INTRODUCTION

The world has witnessed countless changes, especially in the past two decades. One of the most significant aspects of these changes has been technology (Hamari et al., 2014). Technology has completely transformed the way the world operates. Now, people can see and communicate with each other regardless of the distance between them. Things have become faster and more convenient than ever before. One of the most important contributions of technology to the world is the internet (Basten, 2017). The internet has led to the advancement of numerous sectors in society that were previously functioning in traditional ways. Online gaming is a direct result of the internet, while gamification takes it a step further. Unlike online gaming, which involves playing games over the internet, gamification is the concept of incorporating gaming elements into contexts where they are not necessarily required (Seaborn et al., 2015). Gamification is an invention of technology that can be applied to any area of life. Recently, its use has been introduced in the education industry, which ironically is where it is most needed as well (Stieglitz et al., 2017). In recent years, gamification has gained increasing attention and interest in research and education. It is present in our daily lives, often without us even realizing it. Gamification has become established in various areas, including business, the workplace, health, the environment, and particularly the education sector (Klock et al., 2020). In education, gamification is a technology that incorporates game design dynamics into the learning process, inspiring students, engaging with them directly, and significantly enhancing their academic, cognitive, and social skills. It is the method of using systems to attract people, motivate their actions, and encourage learning and problem-solving. This creates a sense of empowerment in the way students work to complete their tasks, making it more attractive and encouraging for the positive value of collaboration, effort, and other play (Saleem et al., 2021). Gamification is a procedure that aims to intensify extrinsic and intrinsic motivation and involve people in tasks through playful activities. Intrinsic motivation is defined as performing activities for their inherent satisfaction, instead of any outside results, while extrinsic motivation is driven by the acquisition of enhancers (Rapp et al., 2019). When designing a gamification strategy, it's important to find the essential motivation to attract users (Aini et al., 2019). This concept is related to self-determination theory (SDT) and has three psychological needs: autonomy (the extent to which a person's actions result from his or her own interests) and ability (ability and sense of self). Achieving tasks at a specific level and building meaningful relationships with others are important aspects of feeling competent and accomplished. Modern-day students no longer find long and tedious lectures effective for understanding concepts. Instead, they prefer to learn through engaging and enjoyable activities, which ultimately leads to better and more effective learning (Fuchs et al., 2014). Limited research on gamification has shown that it is an effective concept for both educators and students, with positive impacts on children's study habits and concept comprehension (Nicholson, 2015). This chapter aims to explore the use of gamification in management education (ME), focusing on how it enhances students' understanding of management concepts.

7.1.1 HIGHLIGHTS OF THE CHAPTER

The chapter aims to highlight the following aspects of gamification:

- What is gamification?
- How can gamification be implemented in the education industry?
- The use and importance of gamification will primarily be analyzed in the management education industry.
- The various constructs and implications of gamification in management education are highlighted in order to do so.
- Lastly, certain real-life examples of companies using gamification in education are considered.

7.2 LITERATURE REVIEW

7.2.1 CONCEPT OF GAMIFICATION

Gamification simply refers to the use of games in a non-gaming context (Ceker et al., 2017). Initially, this concept of gamification was used in the marketing sector. However, with time and the advancement in technology, gamification has also entered areas like health, communication, and other important sectors of society (Caponetto et al., 2017). In this concept of gamification, different aspects of gaming are used in contexts where gaming would not traditionally be used (Arnold, 2014). Even though gamification has become quite common in different areas, there is still a lot of research being conducted on how its implementation can be improved and how it can be implemented in various sectors (Faiella et al., 2015). The main aim of gamification is to ensure that users receive the required support and encouragement to become fully engaged in a given task (Brigham, 2015). When gamification is used in an educational context, it can have extremely positive effects on the way students learn. Gamification can enhance students' interest in areas where they may feel they have not achieved much (Felker et al., 2014). When it comes to consolidating gaming and education, people, in general, are not very positive about this idea. This is mainly because gamification is often associated with just playing games and wasting time, which could be used in a much more productive way (Scepanovik et al., 2015). However, many authors believe that learning in modern times does not have to be dull. With the right kind of technology and tools, gamification has now become an impeccable source of enhancing students' focus and interest in their long-term learning (Huotari et al., 2017).

7.2.2 IMPLEMENTATION OF GAMIFICATION IN EDUCATION

Over the years, scholars have tried to come up with methods through which they could make education more interesting for pupils (Kiryakova et al., 2014). Gaming has been used as an attempt to provide a breakthrough for this particular research. The main aim of using gamification in education is to provide students with activities and tasks that engage them in their studies (Pektas et al., 2017). The mechanism through which these games are designed enables students to develop a competitive spirit, a greater level of interest, and motivation to study effectively (Kirillov et al., 2016). However, sometimes, the use of gamification in education can lead to negative game-based behaviors, which can pose a significant challenge in its implementation (Dicheva et al., 2015). Therefore, the use of gamification should have a clear purpose and reward that it can offer to students. By doing so, the positivity and effectiveness of this resource can prove to be enriching for the progress of these students and for the satisfaction that they require for academic achievement (Buzko et al., 2018). The games that have been created and tested for use in education have been tuned to provide rewards and badges to people who perform the best. In an experiment, it was observed that the students who made use of game-based resources were seen to have a more confident and content approach to the performance that they exhibited (Cahyani, 2016). Literature on gamification has exhibited the need to create games, especially in terms of education, in such a way that they can fulfill the needs of both the student as well as the instructor (Poondej, 2019). This means that through these games, the instructor must be able to create appropriate challenges for the students and that they are eventually able to enhance the difficulty quotient as new knowledge and skill is acquired (Rahardja et al., 2019). Hence, the use of gamification in education is not only attributed to providing students with fun and leisure, but also a complex inclusion and has to be done with utmost care to not defeat the purpose of the concept (Khalil et al., 2018).

7.2.3 GAMIFICATION IN MANAGEMENT EDUCATION (ME)

Board games have always been a way to teach students about complex topics that they will encounter later in life (Silva et al., 2019). Monopoly is an example of the first-ever game used to teach students about finance, assets, and liabilities, which are crucial aspects of management (Muller et al., 2015). During World War 2, mainframe computers were introduced, and a simulation called Top Management was developed (Langendahl, 2017). This tool served two purposes: backing up data related to business studies and providing a gaming experience for students (Urh, 2015). Accounting was observed to be one of the most complex topics for students in management (Villagrasa et al., 2014). To help students better understand these concepts, instructors started using creative technologies, such as digital versions of Monopoly, spreadsheets, and word processors (Putz et al., 2020). These tools served as aids to teaching and learning. Not only does gamification make the subject matter interesting, but it also allows students to revise and practice concepts without the arduous process of traditional learning methods (Leal et al., 2018). In the present day, gamification is being used in various fields such as marketing, tourism, administration, and more to enhance the effectiveness of learning for students in these areas (Ofusu, 2020). Among these sources, the concept is most extensively utilized in education, particularly in ME where theoretical concepts need to be applied practically (Roth et al., 2015). As a result, gamification has become a favorite tool for educators to create a comprehensive and engaging learning environment in the classroom.

7.3 RESEARCH METHODOLOGY

Research on the use of gamification in ME has been extensively conducted in recent years. This study utilized secondary sources from the internet, primarily journal articles from platforms such as EBSCO, Web of Science, Emerald, and Scopus. Two case studies of brands were examined to shed light on how gamification has established a significant presence in the digital age. The study explores the application of gamification in the context of ME.

7.4 VARIOUS CONSTRUCTS THAT MAKE UP GAMIFICATION

In the educational context, several different constructs contribute to the need for gamification in this area (Floryan et al., 2019). These constructs can be defined as student motivation, the flow of studies, attitudes of students, and the progress of learning.

7.4.1 STUDENT MOTIVATION

Motivation is said to be the factor that enables an individual to take the necessary steps towards their goals without being pushed by an outside source (Putz et al., 2015). This motivation is considered to be an important element in the education industry and primarily explains the effort and attention that students put into what they are learning (Sun et al., 2018). Gamification can hence be used to improve the motivation of these students and raise it as and when needed.

7.4.2 FLOW OF STUDIES

The flow of studies refers to the ability of a student to concentrate on a given task for the long term. It is seen that by making use of gamification in education, the flow of what they learn can be improved (Ofusu et al., 2020). By using gaming elements for serious topics, the students would be able to pay more attention to every detail. This would eventually enhance the efficiency through which these learners learn and perform.

7.4.3 STUDENT ATTITUDES

Student attitudes refer to the behaviors and mindset that students have towards their academics. This attitude can be both positive and negative depending on the setting in which these students perform (Zeng, 2016). Hence, it is important to understand that the more positive and interesting a learning environment is, the more positive the attitudes of the students towards their studies would be (Liu et al., 2018). It has been observed that the concept of gamification can have a very positive impact on the attitudes of these students as they get to learn things in a rather fun way.

7.4.4 LEARNING PROGRESS

The progress of students' learning is an important factor in determining whether they will achieve a successful grade or not (Smith, 2017). This progress is often slow and uneven when students are forced to learn in a dull environment. However, this progress becomes faster as soon as an interesting element is introduced (Dorn et al., 2021). This phenomenon has also been observed in gamification. It is believed that by incorporating gamification in education, the progress of students can be significantly improved.

7.5 CONCEPTUAL FRAMEWORK

The studies have shown that gamification in ME is dependent on four major variables: student motivation, learning progress, flow of studies, and student attitudes. Gamification has changed the perspective of ME and the scenario of a teaching–learning experience as it caters to major (Figure 7.1).

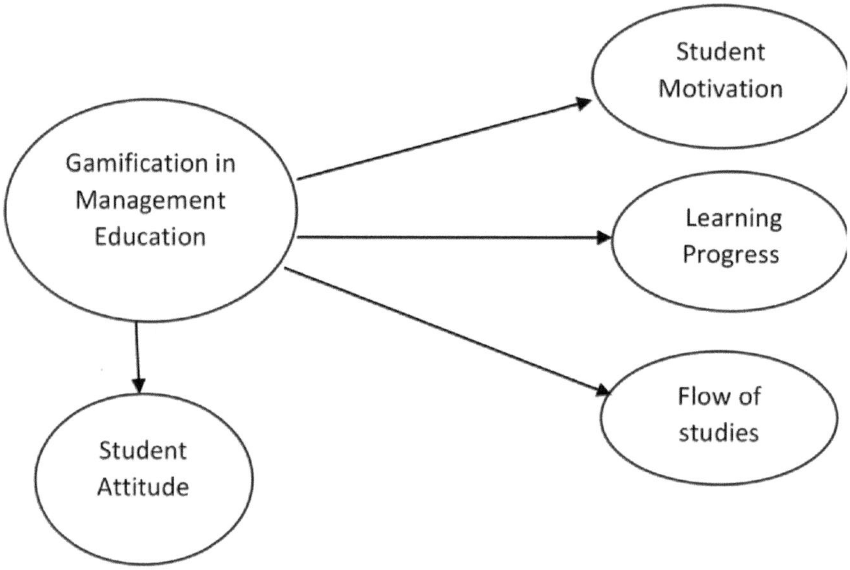

FIGURE 7.1 Conceptual framework.
Source: Authors' compilation.

7.6 CASE STUDIES

7.6.1 IBM

IBM is one of the largest tech organizations in the world. They have been performing exceptionally well, despite the various challenges in the business (Damsa et al., 2016). They have always stayed up to date with their technologies and have ensured to provide their users with the best services possible (Nah et al., 2014). Therefore, there is no doubt that they have also incorporated gamification into their organization in some way or another (Kim, 2015). Since the chapter discusses the use of gamification in ME, the case of IBM will be examined to understand whether the firm has implemented gamification in this context or not.

The area where IBM is seen to be using gamification is in the IBM Case Manager. The major reason why this concept has been used in this particular context is that most of the business that IBM does relies on caseworkers, which is why the managers of the firm want them to work effectively (Joy et al., 2017). Hence, this gamification would help the firm in measuring the task performance well and would also help each of them to see how they are performing in terms of each other so that a kind of healthy competition can be built in the firm. While planning to use gamification in the IBM Case Manager, the management wanted the software that is developed to be such that it can easily align with the priorities of both what the management wants as well as the requirements of the caseworkers (Woodcock et al., 2018). Through this gamification, they wanted to design a user experience where they would eventually be able to reward the employees in proportion to the efforts that they make in their firm and the work that the knowledge and skills that they enhance during the process. To make these features of gamification simple and hassle-free in the firm, the management and the IT team decided to use the IBM Bluemix. The main advantage behind using this was that it did not require any backend development and the entire game could have been created through point and click. Using this tool for incorporating gamification has been a huge benefit for the firm in the sense that the users did not find it difficult to operate this part of the gaming element.

Another aspect of motivational design on which gamification would be used in the IBM Case Manager is social proof. According to this factor of motivation, people are influenced when they observe the behaviors of others. In a gaming context, this means understanding one's own progress compared to others. In the case of the Bluemix gamification application, it would include a Leaderboard API feature, allowing participants to see their rankings. This would be helpful as direct observation of others is often not possible. With the Leaderboard API, caseworkers would still be able to track their performance at work. Additionally, the Leaderboard would only display the performance of the bottom 20% of the group, motivating them to improve and move out of that category.

The last motivation that the caseworkers were given through this gamification concept was rewards. Here, the firm has realized that the rewards that the caseworkers should receive should be sufficient to encourage them to work harder and achieve their goals. However, it was also important to consider the complexity of each task assigned to them. By assessing the complexity of the tasks, the management would be able to determine the appropriate rewards for each caseworker.

The aforementioned strategy of using gamification appears to be highly effective for IBM. They have developed a comprehensive plan to understand the needs of their employees and provide them with the motivation to fulfill those needs. This implementation of gamification within the organization will ultimately help improve the performance of the employees, enabling them to achieve both their personal and professional goals.

7.6.2 DELOITTE

In the past decade, gamification has evolved greatly. Initially, this concept was only seen in places where the risk of including it was low. It started with the marketing sector as it was easy to remove it if the consumers did not like it (Lawanda et al., 2018). However, slowly, and gradually, it became quite popular among people and industries. In the pursuit of making things interesting, more and more companies wanted to include gamification in their operations. While most firms wanted to use gamification for marketing their products and services effectively, firms like Deloitte wanted to use gamification to better educate their employees in management (Alcivar et al., 2016). The senior executives of the firm hold a lot of importance and responsibilities. Hence, to ensure that they had the right kind of skills and knowledge to manage this role well, the management decided to come up with a way through which they could train them again (Dramac, 2016). However, they did not want to spare any extra hours for this training and wanted these executives to learn these skills and techniques while they were on the job. Along with this, since Deloitte has a huge network of consultants working around the world, the management wanted to connect all of these consultants so that sharing information became easier and the employees could also share valuable information through the same network (Sharma, 2014).

Deloitte utilized their own social media platform called Yammer, along with their own geographical location system called Who, What, and Where, for this purpose. Additionally, they employed Badgeville gamification to develop a mobile application that rewarded registered consultants for checking in. The primary objective of this mobile application was to keep consultants informed about each other's whereabouts, fostering a sense of connection, and facilitating the sharing of information. This level of management would have been challenging in a regular context, but the consultants' motivation to receive rewards for checking in made them more inclined to use the application regularly and check in as needed.

Another way through which Deloitte utilized Badgeville gamification in their operations was in the Leadership Academy. The Leadership Academy was an executive training program created to train over 50,000 executives associated with the organization. This program was developed to address the significant challenge of training busy executives. To achieve this, they implemented Badgeville gamification, incorporating various missions, rewards, ranks, and statuses on the website. These features contributed to a 46.6% increase in daily engagement of executives on the website, and a 36.3% increase in weekly engagement.

In both of these cases, Deloitte employed gamification to enhance the activity and engagement of its executives and consultants. This example demonstrates that education is not limited to schools and universities. In the workplace, where individuals apply their knowledge and skills, there is ample opportunity for improvement and enhanced learning. However, similar to higher education (HE), training and learning in organizations do not have to be dull or boring. The introduction of gamification provided a refreshing change to the already demanding lives of personnel. The increased engagement levels observed among executives using the application clearly indicate the success of gamification in this context. This case serves as a lesson on how a simple concept like gamification can be effectively applied in the complex environment of an organization, embedding concrete learning in ME.

7.7 FUTURE OF GAMIFICATION

In a sector like education, improvement and advancement are constant. Without both of these factors, this invaluable sector of the economy would become stagnant, which would, in turn, have an impact on the future generations of that society. The introduction of gamification in this sector has proved to be beneficial in the past years. However, this kind of improvement in this sector cannot end here. This concept of gamification, or making use of gaming elements, would have to be further improved to meet the requirements of the changes that are going to take place in this industry in the future. Along with this, since the use of gamification has proved to be very effective in every sector up until now, it is believed that this concept is therefore here to stay for a long time.

Day by day, education is becoming more complex, but at the same time, it is important for students to perform well and give their best so that they can attain the material success that they have always wanted in life. In such a competitive scenario, gamification comes in as a boon. Talking in terms of education, management has now become one of the leading choices of students all across the globe. However, the mode of attaining a degree in this field can vary. Hence, for the students whose journey of getting a management degree is not smooth, gamification would come in handy. This is because it would enable the learners to not only learn at their own pace but also understand the subject much more effectively than in a traditional classroom.

As it is, the COVID-19 pandemic has completely changed the way we live and learn. Hence, the concept of gamification could eventually become a new way of learning because of the effectiveness and clarity that it brings to the learners. Gamification is indeed an extremely interesting concept, and it could prove to be a game-changer in the field of ME.

7.8 KEY TAKEAWAYS

- Gamification has become an integral part of the education industry and has brought transformation in management education (ME).
- Technology has opened doors for advancements and enhanced engagement, connectivity, and interactivity in ME. As a result, gamification can be used effectively and has become result-driven and extremely progressive.
- Gamification has already been used in sectors like marketing but is now being implemented to enhance the quality of ME.
- By utilizing gamification in their operations, organizations such as Deloitte and IBM have observed positive and commendable changes in both their company and customer operations.
- Gamification has become one of the most effective ways through which people learn, as a result of which, their retention power for the subjects in management has increased.
- Students can become more engaged with their studies through the use of gamification. This approach helps them grasp and comprehend concepts that were once difficult and left unexplained. It provides a solid understanding of management at its core.
- Hence, it is a game-changing intervention in the field of education.

7.9 CONCLUSION

Education and learning have always been two integral parts of society since time immemorial. Without both of these factors, a society can crumble completely, which is why it is important for the methods and quality of education in a particular country to be impeccable. This is only possible when the educational sector is linked with the best that technology has to offer. Technology plays a huge role in the way people learn and grow. If the methods of learning can adapt to the advancements in technology, then learning can become significantly more effective. This is what was expected from the use of gamification in education, specifically in ME. It has been concluded that gamification has already had a positive impact in sectors like marketing and communication. However, its role in ME was yet to be established. Hence, for this chapter, the cases of Course Hero and Deloitte were examined to demonstrate how the utilization of gamification in their operations changed the way students reacted to the content. The level of engagement and retention power of the students increased, and the results were extremely positive in both cases. Therefore, it can be concluded that gamification has proved to be a life-changing invention and has been very beneficial for the ME industry.

KEYWORDS

- **COVID-19 pandemic**
- **gamification**
- **hybrid teaching tools**
- **management education**
- **marketing and communication**
- **student engagement**
- **teaching–learning experience**
- **traditional classroom**

REFERENCES

Aini, Q., Hariguna, T., Putra, P. O. H., & Rahardja, U., (2019). Understanding how gamification influences behavior in education. *International Journal of Advanced Trends in Computer Science and Engineering*, *8*(1.5 Special Issue), 269–274.

Alcivar, I., & Abad, A. G., (2016). Design and evaluation of a gamified system for ERP training. *Computers in Human Behavior*, *58*, 109–118.

Arnold, B. J., (2014). Gamification in education. *Proceedings of the American Society of Business and Behavioral Sciences*, *21*(1), 32–39.

Basten, D., (2017). Gamification. *IEEE Software*, *34*(05), 76–81.

Benito-Santos, A., Dorn, A., Gómez, A. G. L., Palfinger, T., Sánchez, R. T., & Wandl-Vogt, E., (2021). Playing design: A case study on applying gamification to construct a serious game with youngsters at social risk. *Journal on Computing and Cultural Heritage (JOCCH)*, *14*(2), 1–19.

Brigham, T. J., (2015). An introduction to gamification: Adding game elements for engagement. *Medical Reference Services Quarterly*, *34*(4), 471–480.

Cahyani, A. D., (2016). Gamification approach to enhance students engagement in studying language course. In: *MATEC Web of Conferences* (Vol. 58, p. 03006). EDP Sciences.

Caponetto, I., Earp, J., & Ott, M., (2014). Gamification and education: A literature review. In: *European Conference on Games Based Learning* (Vol. 1, p. 50). Academic Conferences International Limited.

Çeker, E., & Özdaml, F., (2017). What "gamification" is and what it's not. *European Journal of Contemporary Education*, *6*(2), 221–228.

Damsa, A., & Fromann, R., (2016). Gamification and gameful approaches in education, business, and IT. *Informatika*, *18*(1), 28–33.

Dicheva, D., & Dichev, C., (2015). Gamification in education: Where are we in 2015? In: *E-Learn: World Conference on E-learning in Corporate, Government, Healthcare, and Higher Education* (pp. 1445–1454). Association for the Advancement of Computing in Education (AACE).

Drmač, T., (2016). *Gamification for Business Practice*. Doctoral dissertation, University of Zagreb. Faculty of Economics and Business. Department of Informatics.

Faiella, F., & Ricciardi, M., (2015). Gamification and learning: A review of issues and research. *Journal of e-Learning and Knowledge Society*, *11*(3).

Felker, K., & Phetteplace, E., (2014). Gamification in libraries: The state of the art. *Reference and User Services Quarterly*, *54*(2), 19–23.

Floryan, M. R., Ritterband, L. M., & Chow, P. I., (2019). Principles of gamification for internet interventions. *Translational Behavioral Medicine*, *9*(6), 1131–1138.

Fuchs, M., Fizek, S., Ruffino, P., & Schrape, N., (2014). *Rethinking Gamification* (p. 344). meson press.

Hamari, J., Koivisto, J., & Sarsa, H., (2014). Does gamification work?—A literature review of empirical studies on gamification. In: *2014 47th Hawaii International Conference on System Sciences* (pp. 3025–3034). IEEE.

Huotari, K., & Hamari, J., (2017). A definition for gamification: Anchoring gamification in the service marketing literature. *Electronic Markets*, *27*(1), 21–31.

Joy, M. M., & Chiramel, M. J., (2017). *Fun is the Future: A Collection of Compelling Gamification Success Stories*. Educreation Publishing.

Khalil, M., Wong, J., De Koning, B., Ebner, M., & Paas, F., (2018). Gamification in MOOCs: A review of the state of the art. In: *2018 IEEE Global Engineering Education Conference (EDUCON)* (pp. 1629–1638). IEEE.

Kim, B., (2015). Gamification in education and libraries. *Library Technology Reports*, *51*(2), 20–28.

Kirillov, A. V., Vinichenko, M. V., Melnichuk, A. V., Melnichuk, Y. A., & Vinogradova, M. V., (2016). Improvement in the learning environment through gamification of the educational process. *International Electronic Journal of Mathematics Education*, *11*(7), 2071–2085.

Kiryakova, G., Angelova, N., & Yordanova, L., (2014). Gamification in education. *Proceedings of 9th International Balkan Education and Science Conference.*

Klock, A. C. T., Gasparini, I., Pimenta, M. S., & Hamari, J., (2020). Tailored gamification: A review of literature. *International Journal of Human-Computer Studies*, *144*, 102495.

Langendahl, P. A., Cook, M., & Mark-Herbert, C., (2017). Exploring gamification in management education for sustainable development. *Creative Education*, *8*(14), 2243–2257.

Lawanda, N., Mohile, R., & Sagarika, D. S., (2018). Gamification and employee engagement: Theoretical review on the role of HR. In: *International Conference on Digital Innovation: Meeting the Business Challenges.*

Leal, C., Silva, R., & Rodrigues, R., (2018). Gamification in management education: Systematic literature review and a proposal of a conceptual research model. In: *ECGBL 2018 12th European Conference on Game-Based Learning* (p. 335). Academic Conferences and publishing limited.

Liu, S., & Idris, M. Z., (2018). Constructing a framework of user experience for museum based on gamification and service design. In: *MATEC Web of Conferences* (Vol. 176, p. 04007). EDP Sciences.

Müller, B. C., Reise, C., & Seliger, G., (2015). Gamification in factory management education–a case study with Lego Mindstorms. *Procedia CIRP*, *26*, 121–126.

Nah, F. F. H., Zeng, Q., Telaprolu, V. R., Ayyappa, A. P., & Eschenbrenner, B., (2014). Gamification of education: A review of literature. In: *International Conference on HCI in Business* (pp. 401–409). Springer, Cham.

Nicholson, S., (2015). A recipe for meaningful gamification. In: *Gamification in Education and Business* (pp. 1–20). Springer, Cham.

Ofosu-Ampong, K., (2020). The shift to gamification in education: A review on dominant issues. *Journal of Educational Technology Systems*, *49*(1), 113–137.

Ofosu-Ampong, K., Boateng, R., Anning-Dorson, T., & Kolog, E. A., (2020). Are we ready for gamification? An exploratory analysis in a developing country. *Education and Information Technologies*, *25*(3), 1723–1742.

Pektas, M., & Kepceoglu, I., (2019). What do prospective teachers think about educational gamification? *Science Education International*, *30*(1), 65–74.

Poondej, C., & Lerdpornkulrat, T., (2019). Gamification in e-learning: A moodle implementation and its effect on student engagement and performance. *Interactive Technology and Smart Education.*

Putz, L. M., & Treiblmaier, H., (2015). *Creating a Theory-Based Research Agenda for Gamification.*

Putz, L. M., Hofbauer, F., & Treiblmaier, H., (2020). Can gamification help to improve education? Findings from a longitudinal study. *Computers in Human Behavior, 110*, 106392.

Rahardja, U., Aini, Q., Graha, Y. I., & Khoirunisa, A., (2019). Implementation of gamification into management of education for motivating learners. In: *Proceeding Interuniversity Forum for Strengthening Academic Competency* (Vol. 1, No. 1, p. 209).

Rapp, A., Hopfgartner, F., Hamari, J., Linehan, C., & Cena, F., (2019). Strengthening gamification studies: Current trends and future opportunities of gamification research. *International Journal of Human-Computer Studies, 127*, 1–6.

Roth, S., Schneckenberg, D., & Tsai, C. W., (2015). The ludic drive as innovation driver: Introduction to the gamification of innovation. *Creativity and Innovation Management, 24*(2), 300–306.

Saleem, A. N., Noori, N. M., & Ozdamli, F., (2021). Gamification applications in e-learning: A literature review. *Technology, Knowledge, and Learning*, 1–21.

Šćepanović, S., Žarić, N. A. Đ. A., & Matijević, T., (2015). Gamification in higher education learning–state of the art, challenges, and opportunities. In: *The Sixth International Conference on e-Learning (eLearning-2015)* (pp. 24, 25).

Seaborn, K., & Fels, D. I., (2015). Gamification in theory and action: A survey. *International Journal of Human-Computer Studies, 74*, 14–31.

Sharma, L., (2014). Gamification-the new tool in learning & development. *SAMVAD, 7*, 51–56.

Silva, R. J. R. D., Rodrigues, R. G., & Leal, C. T. P., (2019). Gamification in management education: A systematic literature review. *BAR-Brazilian Administration Review, 16*.

Smith, T., (2017). Gamified modules for an introductory statistics course and their impact on attitudes and learning. *Simulation & Gaming, 48*(6), 832–854.

Stieglitz, S., Lattemann, C., Robra-Bissantz, S., Zarnekow, R., & Brockmann, T., (2017). *Gamification*. Berlin: Springer.

Sun, J. C. Y., & Hsieh, P. H., (2018). Application of a gamified interactive response system to enhance the intrinsic and extrinsic motivation, student engagement, and attention of English learners. *Journal of Educational Technology & Society, 21*(3), 104–116.

Urh, M., Vukovic, G., & Jereb, E., (2015). The model for introduction of gamification into e-learning in higher education. *Procedia-Social and Behavioral Sciences, 197*, 388–397.

Villagrasa, S., Fonseca, D., Redondo, E., & Duran, J., (2014). Teaching case of gamification and visual technologies for education. *Journal of Cases on Information Technology (JCIT), 16*(4), 38–57.

Wanick, V., & Bui, H., (2019). Gamification in management: A systematic review and research directions. *International Journal of Serious Games, 6*(2), 57–74.

Woodcock, J., & Johnson, M. R., (2018). Gamification: What it is, and how to fight it. *The Sociological Review, 66*(3), 542–558.

Zeng, H., (2016). Gamification study scenario design in MOOCs. *DEStech Transactions on Computer Science and Engineering, (ICTE)*.

Бузько, В., Бонк, А., & Тронь, В., (2018). Implementation of gamification and elements of augmented reality during the binary lessons in a secondary school. *Освітній Вимір, 51*, 74–83.

CHAPTER 8

Is Employability for Management Students Enough? Students' Perspectives on Employability Issues

RENU BALA,[1] SULTAN SINGH,[2] and JOHN EVAR STRID[3]

[1]*Assistant Professor, Jan Nayak Ch. Devi Lal Vidyapeeth, Sirsa, Haryana, India*

[2]*Professor, MMIM Maharishi Markandeshwar (Deemed to be University), Mullana, Ambala, Haryana, India*

[3]*Associate Professor, Northern Illinois University, Oak Park, Illinois, United States*

ABSTRACT

The issue of graduate students' unemployability has become a significant concern for society as a whole. Many students struggle to find good jobs after completing higher degrees. This study focuses on identifying the factors that contribute to the low employability of management graduates. A sample of 350 respondents was taken from various educational institutions, including Central Universities, State Universities, Private Universities, and IIMs. The collected data were analyzed using ANOVA, Factor, and Regression analysis. The results indicate that unfair selection procedures are the most critical factor responsible for the low employability of management students, according to the views of management graduates. This is followed by a lack of industry exposure, skills-based education, a mismatch between required and available skills, and unclear career goals.

Interdisciplinary Approaches in Management Education.
Tanusree Chakraborty, Ramandeep Saini, Rajiv Khosla, Nandita Mishra, and Ashok Natarajan (Eds.)

The study highlights these factors and suggests that educational institutions should identify areas where they lack in providing the necessary knowledge to students and take steps to address these shortcomings. By considering the perspectives of students regarding the factors affecting employability, teachers and educational institutions can improve their services. The study recommends that higher educational institutions make efforts to provide the latest technology equipment and foster academic-industry interfaces to enhance the practical knowledge of management graduates in a real working environment.

8.1 INTRODUCTION

At present, there is a trend of jobless growth, which means a situation where no new jobs are created in net terms, but the country experiences a positive economic growth rate. This happens because more machines are used in the workplace rather than human labor (Horn, 2006). In this era of technology, students who do not possess ICT skills face many challenges, as it has become mandatory to acquire ICT skills to improve the employment situation (Garrido et al., 2010). Employers currently demand qualified students with higher competitiveness and modern employability skills (Fajaryati et al., 2020). In the 20th century, people understood the actual difference between employable and unemployable individuals based on their living standards (Noori & Azmi, 2021). However, freshers work hard on their behalf, and their work-oriented focus increases the linkage impacts of employability skills on organizational performance, resulting in outcomes such as workers' wage gains and career mobility, as recognized by employers (Sung et al., 2013). However, the challenging economic situation indicates that subject-specific knowledge alone is not sufficient for students, as employers now expect them to have multi-skills. Therefore, graduates should focus on developing these skills, which will help them enhance their employability (Fallows & Steven, 2000). Students should make efforts to cultivate the necessary skills in their personality (Bala & Singh, 2020). Additionally, educational institutions should explore new approaches to meet these evolving requirements (Weligamage, 2014). By analyzing market trends, these institutions should strive to prepare students for future job opportunities (Romgens et al., 2019). It is evident that the demand for employability skills is reshaping the landscape of education (Jackson, 2014). Universities are now placing greater emphasis on fostering employability skills in students (Healy et al., 2022). In terms of academic progress, it is crucial to develop various employability skills in students, such as effective communication, foreign language proficiency, teamwork, and efficient learning (Atkins, 1999).

- **Practices in Management Education (ME):** ME plays a vital role in the development of employability skills in students. In such education, there is a wide range of different disciplines. Students can choose a particular stream from a wide range, such as international business, marketing, finance, human resource management, general management, etc. ME offers interdisciplinary practices and career opportunities. It also provides a strong foundation for the IT sector, as there is a major specialization of IT in management studies. The extent to which students can exploit these opportunities depends on their strengths and capabilities.

8.1.1 HIGHLIGHTS OF THE CHAPTER

- Review of literature;
- Research objectives;
- Research hypotheses;
- Data and methodology;
- Results and discussions;
- Conclusion and recommendations;
- Limitations and further scope of the study.

8.2 REVIEW OF LITERATURE

Cassidy (2006) presented the employment issues related to graduates. The study collected data from undergraduate students through peer assessment exercises. It was found that students believe teachers should help them develop skills to overcome entry-level challenges in their careers. Pool (2007) provided a concise definition of employability and suggested ways to understand its various elements. Pan & Lee (2011) explored students' perspectives on employability skills, collecting data from management graduates in Taiwan universities. Regression analysis was used to analyze the data, revealing that students prioritize skills such as applying theory to work, computer literacy, foreign language proficiency, and adaptability.

Rosenberg et al. (2012) examined students' views on employability skills, collecting data from 343 graduate students. The data were analyzed using regression analysis and ANOVA, showing that students value employability skills. However, they face challenges in acquiring these skills as educational institutions often prioritize theoretical knowledge over employability skills. Zahid et al. (2013) stated the viewpoint of business professionals and management graduates regarding the demanded skills in the marketplace. Two separate questionnaires were designed, one for the students and the other for employers. The study results revealed that there is a minimal gap in the views of MBA graduates and employers. Both students and employers share a similar viewpoint regarding educational institutions. Like students, employers also believe that freshers face difficulties at the entry-level job due to the lack of practical knowledge provided by educational institutions. Rao (2014) emphasized the importance of soft skills in developing employability skills in students. Data were gathered through a questionnaire targeting students, faculty members, recruiters, and directors of institutions. The analyzed data indicated that training and placement officers (TPO) should make efforts to provide suitable employment opportunities to students by focusing on reputed organizations. Rao (2015) demonstrated the significance of soft skills training for students. Data were collected from educators, soft-skills trainers, and TPO to progress with the study. The study results indicate that equal importance should be given to soft skills, along with other core subjects in the course curriculum. These skills empower students to be employable and self-employable. Sunardi & Purnomo et al. (2016) analyzed the perspective of management graduates on employability skills. Data was collected from 219 MBA students and analyzed using factor analysis. The results revealed a significant gap between students' expectations and employers' demands, particularly in problem-solving and self-management skills. Malik & Venkatraman (2017) identified the factors contributing to this skill gap and found a substantial mismatch between market demand and the skills acquired by graduates during their education. Chhinzer & Russo (2018) examined employers' perceptions of graduate students' employability. Data was collected from 122 employers who supervised internships of Canadian university students, as well as from individual students. The gathered data has been analyzed using factor analysis. The analysis revealed that employers prioritize generic skills such as general mental ability, time management, willingness to work, and attitudes or behaviors, among others. However, students often lack these skills due to a lack of proper guidance at the educational level. Singh & Jaykumar (2019) identified the skills required for entry-level jobs in the service sector, specifically for youth employment. To collect the data, three separate questionnaires were administered to faculty members, students, and employers. The data from human resource managers and general managers of five-star hotels were gathered at the employer level. The study suggests that students should choose a reputable organization for their internship to build a better career in the future. Bala & Singh (2020) described that bank officials prioritize modern skills such as ICT, new-media literacy, and sense-making over basic employability skills. Khuraisah & Khalid (2020) emphasized the importance of digital literacy skills for students. The study collected secondary data from 22 research papers, which were then thoroughly analyzed. The findings revealed that digital literacy skills are crucial for employers in every organization. Students who possess such skills receive significant attention from recruiters during the hiring process. Bhatnagar (2021) highlighted the reasons for the low employability of MBA graduates in India. To further investigate, data were collected from existing literature. The analyzed data indicated that, in addition to theoretical knowledge and technical aptitudes, employers also value various soft skills in students. With the increasing digitalization in all aspects of life, academia has witnessed numerous changes (Chakraborty & Gupta, 2019; Chakraborty et al., 2021).

8.3 RESEARCH OBJECTIVES

The present research aims to identify the factors contributing to the low employability of management graduates, based on the perspectives of students from selected universities in Haryana, Punjab, and Chandigarh.

8.4 RESEARCH HYPOTHESES

- **H_{01}:** There is no significant difference in students' perception of selected universities regarding factors responsible for the low employability of management graduates.

8.5 DATA AND METHODOLOGY

For the present study, primary data was gathered through a pre-tested structured questionnaire, while secondary data was collected from blogs, reports, magazines, newspapers, etc. To conduct the present study, under Central Universities – Central University of Haryana, Mahendragarh, Central University of Punjab, Bathinda, and Punjab University, Chandigarh was covered. While under State Universities – Kurukshetra University, Kurukshetra, Maharshi Dayanand University, Rohtak, Guru Jambheswar University of Science and Technology, Hisar, Punjabi University Patiala, Guru Nanak Dev University, Amritsar, Maharaja Ranjit Singh Punjab Technical University, Bathinda was included.

Amity University, Gurugram, Chitakara University, Chandigarh, Chandigarh University, Chandigarh, Lovely Professional University, Jalandhar was covered under the head of Private Universities and Indian institute of management (IIMs) – IIM Rohtak and IIM Amritsar were targeted among the IIMs. A sample of 350 respondents was taken, consisting of 60 students from Central University, 130 students from State Universities, 130 students from private Universities, and 30 students from IIMs. The collected data was then analyzed using ANOVA test to compare the responses of the students from the selected universities, followed by factor and regression analysis (Table 8.1).

8.6 RESULTS AND DISCUSSIONS

Table 8.1 presents the students' viewpoint towards the factors responsible for low employability of management students, categorized by institution. The students of Central Universities focused on the following factors: unfair selection procedure ($\bar{x}$=4.88, σ=0.324), lack of industry exposure ($\bar{x}$=4.70, σ=0.497), ineffective use of latest technology in course curriculum ($\bar{x}$=4.60, σ=0.494), outdated course curriculum ($\bar{x}$=4.55, σ=0.534), and mismatch between the required and available skills ($\bar{x}$=4.53, σ=0.503). On the other hand, the students of State Universities emphasized lack of skills-based education ($\bar{x}$=4.70, σ=0.460), lack of innovation and research in teaching ($\bar{x}$=4.65, σ=0.480), lack of industry exposure ($\bar{x}$=4.64, σ=0.498), mismatch between the required and available skills ($\bar{x}$=4.59, σ=0.538), and absence of academic-industry interface ($\bar{x}$=4.58, σ=0.555). The students of Private Universities gave priority to lack of innovation and research in teaching ($\bar{x}$=4.25, σ=0.750), lack of skills-based education

TABLE 8.1 Institution-Wise Students' Viewpoint Towards Factors Responsible for Low Employability of Management Students

Statements	N	Central Universities		N	State Universities		N	Private Universities		N	IIMs		N	Total		ANOVA	
		$\bar{x}$	σ		$\bar{x}$	σ		$\bar{x}$	σ		$\bar{x}$	Σ		$\bar{x}$	σ	F	Sign.
Lack of self-confidence	60	4.52	0.624	130	4.55	0.515	130	4.16	0.620	30	4.98	0.178	350	3.80	0.822	5.434	0.125
Lack of motivation	60	4.10	0.602	130	4.05	0.740	130	3.77	0.766	30	4.51	0.526	350	4.30	0.768	4.234	0.031*
Lack of clear career goals	60	4.23	0.593	130	4.27	0.765	130	3.81	0.728	30	4.90	0.305	350	4.33	0.745	3.442	0.231
Low academic achievements	60	3.80	0.659	130	3.98	0.742	130	3.62	0.771	30	3.65	0.684	350	4.30	0.794	2.431	0.782
Traditional education system	60	4.50	0.537	130	4.55	0.502	130	3.97	0.715	30	4.87	0.346	350	4.32	0.718	6.453	0.665
Lack of skill-based education	60	4.47	0.503	130	4.70	0.460	130	4.22	0.704	30	4.63	0.490	350	4.41	0.747	5.443	0.431
Mismatch between the required and available skills	60	4.53	0.503	130	4.59	0.538	130	4.18	0.616	30	4.97	0.183	350	4.40	0.718	4.443	0.873
Lack of participation in extra-curricular activities	60	3.82	0.624	130	3.72	0.685	130	3.86	0.702	30	4.50	0.529	350	3.83	0.790	6.768	0.261
Outdated course curriculum	60	4.55	0.534	130	4.42	0.633	130	4.12	0.700	30	4.59	0.499	350	4.25	0.802	4.345	0.572
Ineffective use of latest technology in course curriculum	60	4.60	0.494	130	4.52	0.546	130	4.10	0.692	30	4.73	0.450	350	4.31	0.789	3.458	0.685
Lack of industry exposure	60	4.70	0.497	130	4.64	0.498	130	4.21	0.593	30	4.71	0.462	350	4.42	0.755	4.537	0.765
Absence of academic-industry interface	60	4.50	0.548	130	4.58	0.555	130	4.08	0.654	30	4.64	0.478	350	4.29	0.805	6.786	0.539
Unfair selection procedure	60	4.88	0.324	130	4.34	0.644	130	4.17	0.618	30	4.56	0.508	350	4.60	0.742	3.451	0.342
Lack of quality teachers	60	3.62	0.585	130	3.52	0.764	130	3.56	0.779	30	3.89	0.618	350	3.93	0.852	3.221	0.023*
Lack of innovation and research in teaching	60	4.40	0.588	130	4.65	0.480	130	4.25	0.750	30	4.91	0.276	350	4.26	0.869	6.543	0.543
Poor reputation of educational institution	60	3.42	0.612	130	3.57	0.715	130	3.51	0.781	30	4.49	0.532	350	4.22	0.842	3.432	0.001*

TABLE 8.1 *(Continued)*

Statements	N	Central Universities		N	State Universities		N	Private Universities		N	IIMs		N	Total		ANOVA	
		$\bar{x}$	σ		$\bar{x}$	σ		$\bar{x}$	σ		$\bar{x}$	Σ		$\bar{x}$	σ	F	Sign.
Unrealistic salary demand	60	3.36	0.647	130	3.46	0.782	130	3.46	0.788	30	4.41	0.564	350	3.75	0.872	4.567	0.443
Excessive interest in security benefits	60	3.23	0.686	130	3.23	0.794	130	3.42	0.791	30	4.32	0.579	350	3.71	0.912	4.786	0.663
Geographical immobility	60	3.93	0.607	130	3.87	0.730	130	3.28	0.799	30	4.00	0.602	350	4.22	0.874	3.456	0.036*
Slow growth of the economy	60	4.07	0.686	130	3.82	0.755	130	3.31	0.794	30	4.46	0.549	350	4.30	0.745	2.345	0.231

Note: N: Number of respondents; *: Significant at 5% level.

Source: Primary survey (author's calculation).

($\bar{x}$=4.22, σ=0.704), lack of industry exposure ($\bar{x}$=4.21, σ=0.593), mismatch between the required and available skills ($\bar{x}$=4.18, σ=0.616), and unfair selection procedure ($\bar{x}$=4.17, σ=0.618). However, the students of IIMs focused highly on lack of self-confidence ($\bar{x}$=4.98, σ=0.178), mismatch between the required and available skills ($\bar{x}$=4.97, σ=0.183), lack of innovation and research in teaching ($\bar{x}$=4.91, σ=0.276), lack of clear career goals ($\bar{x}$=4.90, σ=0.305), and traditional education system ($\bar{x}$=4.87, σ=0.346). As a whole, the unfair selection procedure ($\bar{x}$=4.60, σ=0.742) is the most critical factor responsible for the low employability of management students, followed by a lack of industry exposure ($\bar{x}$=4.42, σ=0.755), a lack of skills-based education ($\bar{x}$=4.41, σ=0.747), the mismatch between the required and available skills ($\bar{x}$=4.40, σ=0.718), and a lack of clear career goals ($\bar{x}$=4.33, σ=0.745). Statistically, ANOVA results showed a significant difference in students' viewpoint towards a lack of motivation (p=0.031), a lack of quality teachers (p=0.023), a poor reputation of the educational institution (p=0.001), and geographical immobility (p=0.036) at a 5% level of significance, so the null hypothesis (H_{01}) is rejected.

8.6.1 FACTOR ANALYSIS

To identify the factors responsible for the low employability of management students, factor analysis is used. Factor analysis is a statistical measure used to reduce the initial variables and combine them to produce a broad range of factors based on common variables (Inkova, 2014). The Cronbach's Alpha coefficient is 0.862, indicating the reliability and consistency of the data used for analysis (Nunnaly, 1978). Two tests have been applied for the factor analysis. The Kaiser-Meyer-Olkin (KMO) test examines the suitability of factor analysis, and the calculated value of KMO is 0.821, indicating the appropriateness of factor analysis (Table 8.2).

TABLE 8.2 Coefficients of Sampling Adequacy

Coefficients		**Value**
Kaiser-Meyer-Olkin (KMO)		0.821
Bartlett's test of sphericity	Chi-square	7445.973
	d.f.	190
	Sig.	0.000*

Note: N: Number of respondents; *: Significant at 5% level.

Source: Compiled from primary data (SPSS output).

8.6.2 SCREE PLOT

In the following scree plot, factors that lie above the point of inflection must be retained in factor analysis (Cattell, 1966). The scree plot exhibits that the top four factors have eigenvalues greater than 1. Thus, the graphical output of the scree plot also supports the statistical output of PCA and confirms the extraction of four factors under factor analysis (Figure 8.1 and Table 8.3).

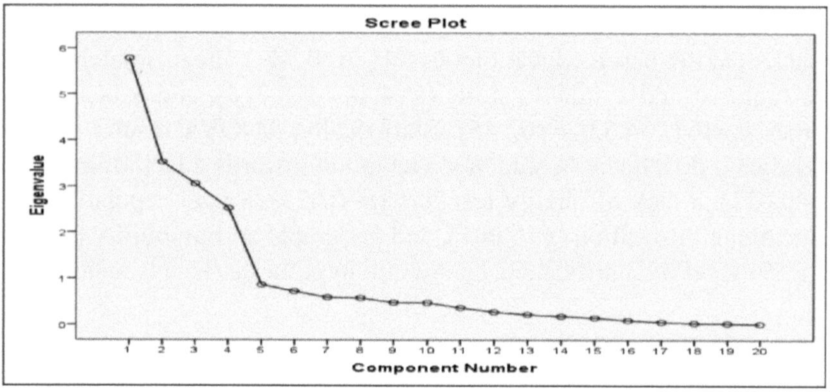

FIGURE 8.1 Scree plot of the variables.

Source: Compiled from primary data (SPSS output).

For the analysis of the Anti-image matrix, only the diagonals with values greater than 0.6 are retained. Measures of sampling adequacy (MSA) help achieve more stable results in factor analysis and identify non-factorable collections of variables (Pathak, 2014).

The factors were derived using the principal component method, which reduces the number of variables while preserving a significant amount of information from the original variables. The 20 variables were divided into four factors. The initial Eigen values represent the total dispersion attributed to each factor. The variance shares of the first, second, third, and fourth factors are 5.783, 3.525, 3.054, and 2.528, respectively. The reliability of the four factors was measured using Cronbach's Alpha coefficient, with values of 0.973, 0.836, 0.968, and 0.947, respectively.

Table 8.5 indicates the commonalities that contribute to the dispersion of variables shared by four factors. The values of communalities range from 0.552 to 0.963, and all values were above the required threshold of

TABLE 8.3 Anti-Image Matrix (Measure of Sampling Adequacy)

Anti-Image	1	2	3	4	5	6	7	8	9	10	11	12	13	14	15	16	17	18	19	20
1	**0.725**[a]	–0.145	0.332	–0.303	0.123	–0.075	–0.037	–0.764	–0.125	0.023	–0.037	0.050	0.060	–0.159	0.075	–0.001	–0.486	–0.339	0.020	–0.091
2	**–0.145**	**0.852**[a]	–0.633	–0.024	–0.107	0.045	0.018	0.147	0.065	0.086	0.062	–0.079	–0.043	–0.021	0.057	–0.064	0.019	0.027	–0.095	0.109
3	0.332	–0.633	**0.697**[a]	–0.685	–0.240	–0.064	–0.062	–0.385	–0.103	–0.037	–0.057	0.092	0.004	0.019	–0.036	0.014	0.000	–0.049	0.076	0.304
4	–0.303	–0.024	–0.685	**0.716**[a]	0.364	0.019	0.041	0.359	0.075	–0.028	0.019	–0.060	0.032	0.033	–0.053	0.088	–0.007	0.011	–0.376	–0.464
5	0.123	–0.107	–0.240	0.364	**0.689**[a]	–0.039	–0.016	–0.134	–0.303	–0.035	–0.020	–0.055	–0.024	0.045	0.049	–0.207	–0.020	–0.002	0.084	–0.895
6	–0.075	0.045	–0.064	0.019	–0.039	**0.887**[a]	–0.148	0.043	–0.049	–0.272	0.056	–0.047	–0.055	–0.080	–0.039	0.011	0.089	0.028	–0.034	0.045
7	–0.037	0.018	–0.062	0.041	–0.016	–0.148	**0.942**[a]	–0.002	0.053	–0.112	–0.092	–0.178	–0.090	–0.043	0.040	0.031	–0.016	0.038	–0.041	–0.026
8	–0.764	0.147	–0.385	0.359	–0.134	0.043	–0.002	**0.768**[a]	0.130	–0.025	0.039	–0.105	–0.045	0.103	–0.045	–0.028	0.004	–0.003	–0.038	0.102
9	–0.125	0.065	–0.103	0.075	–0.303	–0.049	0.053	0.130	**0.919**[a]	0.005	0.041	–0.039	0.060	0.016	–0.036	–0.184	0.004	0.036	–0.034	0.015
10	0.023	0.086	–0.037	–0.028	–0.035	–0.272	–0.112	–0.025	0.005	**0.885**[a]	–0.190	–0.081	–0.226	–0.140	0.080	0.031	–0.042	–0.002	0.017	0.014
11	–0.037	0.062	–0.057	0.019	–0.020	0.056	–0.092	0.039	0.041	–0.190	**0.905**[a]	–0.162	–0.136	–0.077	–0.117	0.023	0.033	–0.032	0.027	–0.010
12	0.050	–0.079	0.092	–0.060	–0.055	–0.047	–0.178	–0.105	–0.039	–0.081	–0.162	**0.905**[a]	–0.192	–0.141	–0.128	0.032	0.065	–0.040	0.023	0.037
13	0.060	–0.043	0.004	0.032	–0.024	–0.055	–0.090	–0.045	0.060	–0.226	–0.136	–0.192	**0.882**[a]	–0.034	–0.300	0.057	–0.111	0.008	–0.019	–0.006
14	–0.159	–0.021	0.019	0.033	0.045	–0.080	–0.043	0.103	0.016	–0.140	–0.077	–0.141	–0.034	**0.807**[a]	–0.293	–0.182	0.110	0.101	–0.026	0.000
15	0.075	0.057	–0.036	–0.053	0.049	–0.039	0.040	–0.045	–0.036	0.080	–0.117	–0.128	–0.300	–0.293	**0.808**[a]	0.018	–0.021	0.031	0.064	–0.025
16	–0.001	–0.064	0.014	0.088	–0.207	0.011	0.031	–0.028	–0.184	0.031	0.023	0.032	0.057	–0.182	0.018	**0.927**[a]	–0.007	0.022	–0.161	–0.005
17	–0.486	0.019	0.000	–0.007	–0.020	0.089	–0.016	0.004	0.004	–0.042	0.033	0.065	–0.111	0.110	–0.021	–0.007	**0.913**[a]	–0.071	0.021	0.013
18	–0.339	0.027	–0.049	0.011	–0.002	0.028	0.038	–0.003	0.036	–0.002	–0.032	–0.040	0.008	0.101	0.031	0.022	–0.071	**0.952**[a]	0.029	0.002
19	0.020	–0.095	0.076	–0.376	0.084	–0.034	–0.041	–0.038	–0.034	0.017	0.027	0.023	–0.019	–0.026	0.064	–0.161	0.021	0.029	**0.922**[a]	–0.013
20	–0.091	0.109	0.304	–0.464	–0.895	0.045	–0.026	0.102	0.015	0.014	–0.010	0.037	–0.006	0.000	–0.025	–0.005	0.013	0.002	–0.013	0.689[a]

[a] Response retained.

Source: Compiled from primary data (SPSS output).

0.5. Therefore, all variables were retained, demonstrating a satisfactory explanation of the identified factors. Table 8.6 displays the rotated component matrix, which shows the relationship between 20 variables formed by the four factors:

- The first factor, labeled 'Factors related to Students,' includes four variables: lack of self-confidence with loadings of 0.962, unrealistic salary demand with loadings of 0.947, lack of participation in extra-curricular activities with loadings of 0.947, and excessive interest in security benefits with loadings of 0.927.
- The second factor, labeled as 'Factors related to Educational Institutions,' includes eight variables. These variables are: unfair selection procedure with loadings of 0.768, absence of academic-industry interface with loadings of 0.734, ineffective use of latest technology in course curriculum with loadings of 0.716, lack of industry exposure with loadings of 0.688, lack of innovation and research in teaching with loadings of 0.671, lack of quality teachers with loadings of 0.666, the mismatch between the required and available skills with loadings of 0.585, and lack of skill-based education with loadings of 0.567.
- The third factor, 'Factors related to Lack of Family Support,' includes four variables: lack of clear career goals with 0.960 loadings, low academic performance, lack of emotional support, and lack of financial support (Tables 8.4–8.6).

  Achievements with 0.959 loadings, lack of motivation with 0.953 loadings, and geographical immobility with 0.905 loadings.
- The fourth factor, 'Factors related to Government Policies,' includes four variables: traditional education system with 0.958 loadings, slow growth of the economy with 0.951 loadings, outdated course curriculum with 0.911 loadings, and poor reputation of the educational institution with 0.879 loadings.

8.6.3 REGRESSION ANALYSIS

The students of the selected educational institution focused on the unfair selection procedure as a factor responsible for the low employability of

TABLE 8.4 Extraction Sums of Squared Loadings and Rotation Sums of Squared Loadings

Components	Initial Eigen Values			Extraction Sum of Squared Loadings			Rotation Sum of Squared Loadings		
	Total	Percentage of Total Variance	Cumulative (Percent)	Total	Percentage of Total Variance	Cumulative (Percent)	Total	Percentage of Total Variance	Cumulative (Percent)
1	5.783	28.914	28.914	5.783	28.914	28.914	3.854	19.270	19.270
2	3.525	17.625	46.540	3.525	17.625	46.540	3.774	18.871	38.141
3	3.054	15.268	61.807	3.054	15.268	61.807	3.732	18.660	56.801
4	2.528	12.638	74.445	2.528	12.638	74.445	3.529	17.644	74.445
5	0.860	4.300	78.745						
6	0.719	3.595	82.340						
7	0.589	2.943	85.283						
8	0.576	2.880	88.163						
9	0.478	2.388	90.550						
10	0.472	2.358	92.908						
11	0.367	1.833	94.741						
12	0.272	1.362	96.103						
13	0.218	1.092	97.194						
14	0.182	0.910	98.104						
15	0.149	0.744	98.848						
16	0.095	0.475	99.323						
17	0.059	0.295	99.617						
18	0.035	0.175	99.793						
19	0.027	0.133	99.926						
20	0.015	0.074	100.000						

Source: Compiled from primary data (SPSS output).

TABLE 8.5 Extraction of the Variables

Sl. No.	Statements	Initial	Extraction
1.	Lack of self-confidence	1.000	0.963
2.	Lack of motivation	1.000	0.936
3.	Lack of clear career goals	1.000	0.954
4.	Low academic achievements	1.000	0.953
5.	Traditional education system	1.000	0.942
6.	Lack of skill-based education	1.000	0.573
7.	The mismatch between the required and available skills	1.000	0.552
8.	Lack of participation in extra-curricular activities	1.000	0.935
9.	Outdated course curriculum	1.000	0.842
10.	Ineffective use of the latest technology in the course curriculum	1.000	0.556
11.	Lack of industry exposure	1.000	0.583
12.	Absence of academic-industry interface	1.000	0.577
13.	Unfair selection procedure	1.000	0.618
14.	Lack of quality teachers	1.000	0.566
15.	Lack of innovation and research in teaching	1.000	0.599
16.	The poor reputation of educational institution	1.000	0.783
17.	Unrealistic salary demand	1.000	0.919
18.	Excessive interest in security benefits	1.000	0.881
19.	Geographical immobility	1.000	0.832
20.	Slow growth of the economy	1.000	0.923

Source: Compiled from primary data (SPSS output).

TABLE 8.6 Factors Labelling and Loading

Factor(s)	Factors Labelling	Statements	Factor Loading			
			I	II	III	IV
Factor I	Factors related to students	Lack of self-confidence	0.962			
		Unrealistic salary demand	0.947			
		Lack of participation in extra-curricular activities	0.947			
		Excessive interest in security benefits	0.927			
Factor II	Factors related to educational institutions	Unfair selection procedure		0.768		
		Absence of academic-industry interface		0.734		

TABLE 8.6 *(Continued)*

Factor(s)	Factors Labelling	Statements	Factor Loading			
			I	II	III	IV
		Ineffective use of latest technology in the course curriculum		0.716		
		Lack of industry exposure		0.688		
		Lack of innovation and research in teaching		0.671		
		Lack of quality teachers		0.666		
		Mismatch between the required and available skills		0.585		
		Lack of skill-based education		0.567		
Factor III	Factors related to lack of family support	Lack of clear career goals			0.960	
		Low academic achievements			0.959	
		Lack of motivation			0.953	
		Geographical immobility			0.905	
Factor IV	Factors related to government policies	Traditional education system				0.958
		Slow growth of economy				0.951
		Outdated course curriculum				0.911
		Poor reputation of educational institution				0.879
Cronbach's Alpha Coefficient of the Factors						
0.973		0.836	0.968	0.947		

Source: Compiled from primary data (SPSS output).

management students, as it has the highest mean score ($\bar{x}$=4.60). Therefore, the choice of the surrogate variable for subsequent analysis depends on the highest mean value. The dependent variable is 'unfair selection procedure.' The independent variables are extracted from factor analysis, namely factors related to students, factors related to educational institutions, factors related to lack of family support, and government policies.

8.6.4 REGRESSION EQUATION

$$Y=a+\beta_1 X_1+\beta_2 X_2+\beta_3 X_3+\beta_4 X_4+e_i$$

where; Y is the unfair selection procedure; α is the Y intercept; β is the slope coefficient; X_1 is the factors related to students; X_2 is the factors related to educational institutions; X_3 is the factors related to lack of family support; X_4 is the factors related to government policies; and e_i is the residual or error term.

Table 8.7 presents the results of the regression analysis. In the summary of model fit, the R-Square value is 0.618, indicating that the model has explained 61% of the total variance explained by the experimental factors, namely factors related to students, educational institutions, lack of family support, and government policies. The remaining 39% of the variance may be attributed to other factors not covered in the study. The Adjusted R-Square value, a modified version of R-Square, is 0.614, which is almost equal to the R-Square value (0.618), suggesting that the additional input variables are not contributing significantly to the model. Factor-I has a beta value of 0.117, followed by t-statistics (04.756) and p-value (0.000).

TABLE 8.7 Model Summary, ANOVA Results, and Coefficients of Factors

Model	**R**	**R-Square**	**Adjusted R-Square**	**Std. Error of Estimate**	
1	0.786	0.618	0.614	0.461	
Model	Sum of Squares	d.f.	Mean Square	F	Sig.
Regression	118.851	004	29.713	139.759	0.000*
Residual	73.347	345	0.213		
Total	192.197	349			
Model	**Unstandardized Coefficients**		**Standardized Coefficients**	**t-Statistic**	**Sig. Value**
	Beta	**Std. Error**	**Beta**		
(Constant)	4.597	0.025	–	186.527	0.000*
Factor I	0.117	0.025	0.158	04.756	0.000*
Factor II	0.570	0.025	0.768	23.077	0.000*
Factor III	0.048	0.025	0.065	01.964	0.000*
Factor IV	0.003	0.025	0.004	0.115	0.000*

Note: *: Significant at 5% level.

Source: Compiled from primary data (SPSS output).

The beta value of Factor-II is 0.570, followed by t-statistics (23.077) and p-value (0.000). Factor-III has a beta value of 0.048, followed by t-statistics (01.964) and p-value (0.000). Factor-IV has a beta value of 0.003, followed by t-statistics (0.115) and p-value (0.000). These results indicate that all four factors, namely factors related to students, educational institutions, lack of family support, and government policies, are significant at a 5% level of significance. Therefore, it can be concluded that all the extracted factors contribute to the low employability of management.

8.7 CONCLUSION AND RECOMMENDATIONS

The outcomes indicate that students face numerous challenges in terms of employability, as various factors hinder the career prospects of graduates. According to students from Central Universities, unfair selection procedures, lack of industry exposure, ineffective use of technology in the curriculum, outdated course content, and a mismatch between required and available skills are the primary factors contributing to the low employability of management graduates. Similarly, students from State Universities emphasized the lack of skills-based education, limited innovation and research in teaching, insufficient industry exposure, a mismatch between required and available skills, and a lack of academic-industry interface. On the other hand, students from Private Universities prioritized the lack of innovation and research in teaching, followed by a lack of skills-based education, limited industry exposure, a mismatch between required and available skills, and unfair selection procedures. However, students from IIMs highlighted the lack of self-confidence, a mismatch between required and available skills, limited innovation and research in teaching, a lack of clear career goals, and the traditional education system. Unfair selection procedures are the primary factor responsible for the low employability of management students as a whole, followed by a lack of industry exposure, skills-based education, a mismatch between required and available skills, and a lack of clear career goals. To address these factors and improve the employability of management graduates, educational institutions should prioritize the use of the latest technological equipment and focus on imparting practical knowledge to students. Additionally, organizing conferences, seminars, educational trips, and other similar activities on behalf of the educational institution can provide students with valuable experience in a real working environment.

8.8 LIMITATIONS AND FURTHER SCOPE OF THE STUDY

The present research shows the perception of management students regarding factors responsible for the low employability of management graduates. The study is limited to a restricted number of universities and areas. Future studies can be extended to other universities and states by including students from other disciplines as well.

KEYWORDS

- **educational institutions**
- **employability**
- **employability issues**
- **factor analysis**
- **management graduates**
- **perspective**
- **regression analysis**

REFERENCES

Atkins, M. J., (1999). Oven-ready and self-basting: Taking stock of employability skills. *Teaching in Higher Education*, *4*(2), 267–280.

Bala, R., & Singh, S., (2020). Employability skills of management students: A study of banker's viewpoint. *Psychology and Education Journal*, *57*(9), 2908–2914.

Bala, R., & Singh, S., (2020). Perception of students towards employability skills. *Shodh Sarita*, *7*(28), 107–112.

Bhatnagar, N., (2021). Employability and skill gap among MBA graduates in India: A literature review. *Industrial and Commercial Training*, *53*(1), 92–104.

Cassidy, S., (2006). Developing employability skills: Peer assessment in higher education. *Education + Training*, *48*(7), 508–517.

Chakraborty, T., & Gupta, D., (2019). Examining dynamics of emotional intelligence and entrepreneurial intentions among management and non-management students. *The International Journal of Indian Psychology*, *7*(2).

Chakraborty, T., Tripathi, M., & Saha, S., (2021). The dynamics of employee relationships in a digitalized workplace: The role of media richness on workplace culture. In: *Critical Issues on Changing Dynamics in Employee Relations and Workforce Diversity* (pp. 175–205). IGI Global

Chhinzer, N., & Russo, A. M., (2018). An exploration of employer perceptions of graduate student employability. *Education + Training, 60*(1), 104–120.

Dacre, P. L., & Sewell, P., (2007). The key to employability: Developing a practical model of graduate employability. *Education + Training, 49*(4), 277–289.

Fallows, S., & Steven, C., (2000). Building employability skills into the higher education curriculum: A university-wide initiative. *Education + Training, 42*(2), 75–83

Garrido, M., Sullivan, J., & Gordon, A., (2010). Understanding the links between ICT skills training and employability: An analytical framework. *ICTD '10: Proceedings of the 4th ACM/IEEE International Conference on Information and Communication Technologies and Development* (pp. 1–10). Article No.: 15.

Healy, M., Hammer, S., et al., (2022). Mapping graduate employability and career development in higher education research: A citation network analysis. *Studies in Higher Education, 47*(40) 799–811.

Horn, G., (2006). Educational solutions to improve the employability of senior high school learners. *South African Journal of Education, 26*(1), 113–128.

Ibidunni, A. S., Mozie, D., & Ayeni, A. W. A. A., (2021). Entrepreneurial characteristics amongst university students: Insights for understanding entrepreneurial intentions amongst youths in a developing economy. *Education + Training, 63*(1), 71–84.

Jackson, D., (2014). Testing a model of undergraduate competence in employability skills and its implications for stakeholders. *Journal of Education and Work, 27*(2), 220–242.

Khuraisah, M. N., et al., (2020). Preparing graduates with digital literacy skills toward fulfilling employability need in 4IR era: A review. *(IJACSA) International Journal of Advanced Computer Science and Applications, 11*(6), 307–316.

Malik, G., & Venkatraman, A., (2017). The great divide: Skill gap between the employer's expectations and skills possessed by employees. *Industrial and Commercial Training, 49*(4), 175–182.

Noori, M., & Azmi, F., (2021). Students perceived employability: A systematic literature search and bibliometric analysis. *Eurasian Journal of Business and Economics, 14*(28), 37–56.

Pan, Y. J., & Lee, L. S., (2011). *Academic Performance and Perceived Employability of Graduate Students in Business and Management an Analysis of Nationwide Graduate Destination Survey* (Vol. 25, No. 1, pp. 91–103). Paper presented at International Conference on Asia Pacific Business Innovation & Technology Management.

Rao, M. S., (2014). Enhancing employability in engineering and management students through soft skills. *Industrial and Commercial Training, 46*(1), 42–48.

Rao, M. S., (2015). Step by step to soft-skills training: How to enhance employability skills in students. *Human Resource Management International Digest, 23*(6), 34–36.

Romgens, I., Scoupe, R., et al., (2019). Unravelling the concept of employability, bringing together research on employability in higher education and the workplace. *Studies in Higher Education, 45*(2), 1–16.

Rosenberg, S., Heimler, R., et al., (2012). Basic employability skills: A triangular design approach. *Education+Training, 54*(1), 7–20.

Singh, A., & Jaykumar, P., (2019). On the road to consensus: Key soft skills required for youth employment in the service sector. *Worldwide Hospitality and Tourism Themes, 11*(1), 10–24.

Sunardi, et al., (2016). Employability skills measurement model's of vocational student. *Proceedings of the International Mechanical Engineering and Engineering Education Conferences (IMEEEC 2016),* AIP Conference Proceedings.

Sung, J., Man, N. M. C., Loke, F., & Ramos, C., (2013). The nature of employability skills: Empirical evidence from Singapore. *International Journal of Training and Development,* 76–193.

Weligamage, S., (2014). Graduates' employability skills: Evidence from literature review. *Enhancing Employability Through Quality Assurance*. Proceedings of University of Kelaniya, Sri Lanka.

Zahid, M. M., Naeem, A., et al., (2013). Business education curriculum and management skills: A viewpoint of students and managers. *Academic Research International Journal, 4*(4), 431–441.

CHAPTER 9

Contemporary Challenges of Management Education in India: Review and Assessment

JASKIRAT SINGH[1] and DHIRAJ SHARMA[2]

[1]*Research Scholar, Punjabi University, Patiala, Punjab, India*

[2]*Faculty Member, School of Management Studies, Punjabi University, Patiala, Punjab, India*

ABSTRACT

Management education (ME) has witnessed a significant transition in the past few years, reflecting globalization as a crucial component of the present times. Apex agencies like the UGC and the AICTE have authorized various management curricula over the past decade. The advancement of the ME system is crucial in understanding the business environment, its challenges, complexities, and other key variables. It broadly covers essential issues such as training to survive in a competitive job market, the need for students to gain knowledge and experience with specialization to increase employment opportunities, the nature of corporate jobs, and the seemingly limitless possibilities offered by technology induction. The present study is based on a thorough literature review focused mainly on the challenges and bottlenecks experienced by ME in India.

9.1 INTRODUCTION

Management education (ME) in India has a rich history that dates back to the pre-independence era. In 1913, the world's first quality management

Interdisciplinary Approaches in Management Education.
Tanusree Chakraborty, Ramandeep Saini, Rajiv Khosla, Nandita Mishra, and Ashok Natarajan (Eds.)

institution was formed in Bombay, with a second college opening its doors to applicants in Delhi in 1920. These business colleges provided teaching in the foundations of barter and commerce to clerks and supervisors from many industries, including banking, transportation, and accounting. In 1950, the Department of Commerce at Andhra University began its first-ever MBA curriculum, making it the first of its type worldwide. The Indian Institute of Social Welfare and Business Management, based in Kolkata, was the country's first business school, opening its doors in 1953. The Both India Institute of Management and Labor Welfare, as well as the Department of Management of Delhi University, were established in 1968, respectively (Balamurugan & Dharsini, 2017; Krishnamoorthy, 2017; French & Grey, 2017; Singh & Jasial, 2021).

ME has seen significant changes over the past 70 years, particularly in India. A substantial upsurge in demand for MBA education occurred during economic liberalization and subsequent fast economic expansion in the 1990s (Chauhan et al., 2021). This demand paved the way for numerous public and private schools to offer MBA curricula around the country. Because the MBA is a highly in-demand credential, it is often seen as a remedy for resolving all business problems (Chakraborty, 2021). In contrast, the proliferation of master's degrees has resulted in systemic faults, which have further raised many problems. The present political and economic turmoil has highlighted flaws in our country's ME and has created a need for immediate corrective measures to address concerns due to the apparent combined effects of digital disruption and the COVID-19 pandemic outbreak (Porter & McKibbin, 1988; Collet & Vives, 2013; Edumate, 2021).

Management is a profession in which Indians are highly sought after. Nowadays, the brightest and sharpest students aspire to pursue Management because it offers a guaranteed path to well-paying professions with excellent development opportunities. In 2008–2009, more than one lakh students were enrolled in programs leading to a Master of Business Administration or its equivalent. However, are management schools adequately supplying qualified candidates to businesses in desperate need of their services? The simple answer is no. Many of the over 2,000 management institutions accredited by AICTE (All India Council for Technical Education) have a small number of academic staff, most of whom lack actual management expertise and limit high-quality research in their fields. Additionally, only 250–300 management schools allow themselves to be evaluated, which would make for an easy evaluation if everyone participated. However, management schools should be at the forefront of ME, especially for older students with prior job experience, but unfortunately, this is not the case currently.

Instead of focusing solely on business schools, it is crucial to enhance other educational streams, such as commerce education and the management orientation of higher studies like accountancy and administrative studies, rather than just B-schools (Singh & Jasial, 2021). These serve as the foundation for the numerous executives in Indian companies who require a strong base to have an early and meaningful impact as managers. However, the most significant challenge faced by Management today is the placement of recent B-school graduates in their desired positions. In 2007, the number of unlisted businesses in India was approximately 12,500, including public sector organizations, banks, financial enterprises, and service companies (Krishnamoorthy, 2017). The top 75 of these businesses are comparable to the best publicly traded corporations. However, the majority of the others have poor performance in terms of sales growth, profit ratios, and overall stability, and it is in these areas that management graduates should be able to make a significant impact (Azevedo & Shane, 2019).

According to the research findings, consultancy companies, multinational corporations, and financial institutions primarily target MBA degree holders. Graduates and postgraduates in economics, finance, and engineering are well-positioned to meet the significant demands of the industry, including infrastructure, property investment, small-scale units, non-banking financial institutions, and educational organizations, among others (Rubens et al., 2020). However, the limited number of management graduates compared to the overall demand creates a sense of superiority and arrogance among many of them. These graduates complete their MBAs in their early 20s and are often flooded with employment offers. Some even receive monthly salaries as high as five lakh rupees. The media, employers, and students rate these institutions as high quality based on the speed of student placements and the size of their compensation packages (Kakabadse & Vinnicombe, 2018).

Consequently, while idealistic thinking should be established early, greed is implanted. Assume that management teams are always coming up with new strategies to enhance stock values, lavishly rewarding themselves in the process, and shifting the costs to society (Roddy et al., 2017). They are merely doing what they were taught in business classroom sessions on strategies: transaction cost economics and agency theory, maximizing the value of a firm, a product, and their own worth. Rather than focusing only on maximizing business profits at any cost, ME should emphasize society and social good (Bhatnagar & Ramachandran, 2021).

The present study focuses on the contemporary challenges of ME in India through available literature. The study is organized into five parts. The first part, the Introduction, discusses the background of ME. This part is followed by the second part, which entails the present scenario of ME in India. The third part lists the strategies followed in ME. The fourth part reviews the extensive literature on current approaches to ME. The final part summarizes the study by listing the key points discussing the 'Problems Prevailing in the Indian Management Education System' and 'Good Practices Adopted in Modern-Day B-Schools in Developed Countries.' Based on these key findings, the 'Conclusion' and 'The Impending Change and The Path Forward' conclude the study.

9.1.1 MANAGEMENT EDUCATION (ME) IN INDIA: PRESENT SCENARIO

In India, ME is available in a variety of formats, all designed to help learners enhance their cognitive abilities and background knowledge while also developing their executive persona and managerial skills. This is achieved through a blend of business acumen and global training that caters to the individual needs of students.

With a degree in Management Sciences, students can enhance their intellectual capabilities by understanding the theoretical and philosophical foundations of the modern profession. In MBA or MS programs, students receive a comprehensive ME that incorporates internationally recognized best practices, while also allowing flexibility to customize the curriculum based on the student's specific entrepreneurial and societal context.

The ME program offers undergraduate, postgraduate, and doctoral levels of study. Students have the option to choose from full-time, part-time, executive, remote learning, and specialty courses. These courses may have different names, but they all aim to provide a well-rounded ME:

- Masters in finance control;
- Master of business administration;

- Master of management studies;
- Master of public administration;
- Postgraduate diploma in financial management;
- Postgraduate diploma in human resource management;
- Postgraduate diploma in management;
- Postgraduate diploma in operations management;
- Postgraduate diploma in public administration.

9.1.2 STRATEGIES OF MANAGEMENT EDUCATION (ME)

Enhanced collaboration between industry and academic institutions is necessary to produce an improved course structure, innovative teaching methods, and increased employment prospects in the 21st century. The following issues must be considered to achieve industry-endorsed employability capacity: mobilizing resources and communication skills, business-related abilities and academic talent, goal variety, and goal selection concerning organizational behavior. Unfortunately, the ability to regulate one's personal time and think critically is given the least consideration. Measures should be taken to enhance these elements for the benefit of students, allowing quality-conscious undergraduates to be admitted to MBA programs and achieve higher performance levels. Surprisingly, despite being considered innovative teaching techniques, management games, case studies, group discussions, industry-sponsored guest lectures, and seminars on research basics are not being conducted. However, it is likely that this approach will be taken into consideration in the near future. In terms of essential features for students' development, wireless Internet access (Wi-Fi), video charts, an electronic library, and ready-made e-content are a few examples of the ICT (information and communication technology) features.

Additionally, there are plans to develop infrastructure of this sort at other colleges as well. Growing faculty should be a part of the management studies program from the very beginning. Members of the teaching staff must maintain solid interpersonal ties with their pupils while also gaining business exposure through consulting assignments (Reddy, 2020; Davenport et al., 2003; Rubens et al., 2020).

9.2 LITERATURE REVIEWS ON CURRENT APPROACHES TO MANAGEMENT EDUCATION (ME)

An extensive compilation of literature was conducted to synthesize data from various sources worldwide on goal-oriented coaching techniques in management research. Utilizing Google Scholar, we searched for open-access national and international journals to ensure comprehensive coverage of the field's literature, with a focus on materials most relevant to the study objective.

9.2.1 CONTEMPORARY MATTERS OF MANAGEMENT EDUCATION (ME) IN INDIA

According to Bartlett et al. (1997); Volery (2000); Fuller (2002); Lindell (2006); and Balamurugan & Dharsini (2017), it is critical to ensure that emergency management curricula include the whole range of challenges that confront professional emergency management personnel. Due to the high competition for positions in emergency management, students who obtain work experience in a professional context will have a better chance of securing a position. The quality of such programs should also be examined. Based on the ideas presented here, the majority of emergency management programs are expected to face various difficulties and opportunities due to the complex nature of the profession and the virtually limitless possibilities offered by technological advancements.

One of the most significant human innovations of the 21st century, according to Canals (2009), has been the substantial development in the scientific-industrial field that has resulted in the creation of a new profession. The development of businesses has become extremely demanding, and it has played and continues to play a noteworthy part in ME. It trains individuals in the management skills needed to handle the most challenging issues faced today. Knowledge and concepts are imparted to professional managers through formal education. The present-day entrepreneurial arrangement has advanced the technical abilities of management, but it still lacks the fundamental components for successfully directing an enterprise.

Based on their research, Roddy et al. (2017); and Ottewill (2003) concluded that the problem of student diversity also necessitates tutors applying their knowledge and skills in the design of their learning, teaching, and assessment techniques. They must include opportunities for all pupils to understand the nature of what they are teaching. Business and management instructors, in particular, bear the obligation of developing and maintaining a positive learning environment for their students. The increasing size and variety of the student population and the relative reduction in the value of the resource unit have placed tremendous pressure on business and management educators to respond effectively to issues that impact all students to a greater or lesser extent.

Mamun & Mohamad (2009); Canals (2009); and Harasim (2006) have stated that ME needs to adapt to the demands of changing environments in order to develop future managers who possess all the necessary skills. The standard curriculum of management schools may not adequately prepare individuals for the challenges they would face as professional managers in today's society. It is crucial for their long-term survival to nurture knowledgeable individuals who are ready to tackle the complexities of the business landscape. The MBA program must be tailored to the requirements of the current business setting.

Azevedo & Shane (2019); Johannesson (2010); and Benson et al. (2013) have emphasized that current trends have an impact on the evolution of ME, and it must adjust its market structures to keep up with the times. Culture encompasses a set of beliefs and attitudes towards change. Due to globalization, change has become a constant process. It is necessary to embrace a political power system that seeks to bring about change and facilitates the adoption of a progressive program at the Institute. The program was founded on the principle of teaching continuous improvement in existing and viable programs and activities, as well as in new ones (Cohn, 2008). The global economic environment has gained new significance. It has transitioned from being a country-specific, state-specific, or city-specific education to an international education. The Institute's reputation and image must be of high quality to familiarize new entrants with the Institute's working system and environment. An intellectually engaging, dynamic, and professional atmosphere is required to retain competent faculty members (Mitra, 2012).

Several researchers, including Benson et al. (2013); and Harasim (2006), believe that online learning (OLL) is becoming a more viable alternative for college-level business education. The facility will necessitate an educational model to assist them in creating their online courses. A pedagogical framework for online educators may be found in the community of inquiry (CoI) model, which emphasizes presence – social, cognitive, and instructional – and places a strong focus on presence.

9.2.2 CHALLENGES OF MANAGEMENT EDUCATION (ME)

Hall et al. (2007); Balamurugan & Dharsini (2017); and Azevedo & Shane (2019) found that a personalized and adaptable curriculum for ME is becoming increasingly popular. It has been shown that a well-designed personalized curriculum provides students with a greater sense of ownership and cooperation, which can result in greater effectiveness compared to a broader curriculum. Adult ME is associated with a customized curriculum, which may be a response to current challenges. Scholars are transitioning to a new era of educational knowledge through the Internet, which facilitates the development of new managerial skills through virtual platforms. Benchmarking is one approach that can be utilized to assist in designing and implementing a tailored MBA program. According to a student satisfaction survey conducted upon graduation, a benchmarked tailored MBA curriculum provided more significant insights and job-related abilities than a generic MBA program.

According to Muniappan (2007); and Thomas & Mengel (2008), Financial Management, economy, business process management, market research, and Management are only a few of the sub-fields within business management. In a global environment, this organization successfully manages cultural and modern management methods and ideas from around the globe; it has proper knowledge of recent and historical models, thoughts, procedures, and guidelines, allowing it to manage effectively and smartly in a global ecosystem.

Managers have a variety of informational responsibilities, including those of monitor, disseminator, and spokesman. Technical skills are the ability to use technical knowledge or experience in a specific situation. Many employees receive on-the-job training and education to further their technical abilities. When it comes to human skills, the capacity to collaborate with, understand, and motivate others, both individually and in groups, is essential. Good human skills communicate and inspire, and technical skills have higher conceptual skills than their counterparts in the field.

In their study, Balamurugan & Dharsini (2017) discovered that ME, particularly among faculty, is at a crossroads in India. A lack of innovation will almost certainly result in the conventional university system of learning and preparing tomorrow's business leaders being outclassed and abandoned in an increasingly diversified and technologically advanced global market. With their inherent complexity, international markets must be incorporated into the primary undergraduate and postgraduate management teaching frameworks, respectively. As a side note, information systems should be endorsed to increase academic effectiveness while also being recognized as a significant contender in the expansion of educational services, which is an instant necessity in order to structure worldwide changes to improve the competitive capacity in quality ME.

According to the findings of their study, Mahajan et al. (2014) state that ME worldwide is currently facing a unique crisis of relevance in today's corporate environment, particularly in the quality of Master of Business Administration (MBA) programs. Youngsters, teaching staff, society, manufacturers, the government, and the global community as a whole believe that business schools need to revitalize ME to meet the aspirations of the Indian and global communities. To achieve this, business schools in India must re-energize ME. In addition to addressing fundamental issues such as shortages of academic staff, lack of governance arrangements, absence of efficient supervisory authorities, substandard research and literature, scarcity of educational innovations, decreased industry-institute functionality, and limited employment prospects within and between business schools, a wide range of counseling can contribute to the development of a comprehensive perspective for effective ME.

Mahajan & Sharma (2012) conducted a study examining the current state of ME in India, as well as the trends and contemporary issues faced by ME in the country. The findings suggest that ME is now more crucial than ever in the "global era." The main challenge for ME techniques is to become more practical and industry-oriented. The credibility of management is constantly at risk due to the disconnect between business schools and the real world, as well as the inability to keep up with the rapid pace of change. Therefore, it is imperative to align ME with global developments in order to enhance competitiveness through overall quality management. To bridge the gap between industry requirements and academic curriculum, ME needs to be comprehensive, targeted, and tailored, with a strong emphasis on corporate awareness, grooming, and the development of managerial skills.

Balamurugan & Dharsini (2017) have highlighted that the globalized world necessitates the development of corporate executives capable of both leadership and management. Traditional functional sectors have altered their path in well-organized ways to evolve and solidify soft skills, state-of-the-art self-learning, and an awareness of the organizational environment, both internally and externally, from their previous trajectory. To face the current challenges of managing a firm worldwide, business schools are redesigning and reshaping their ME programs.

In the opinion of Child (1996); and Dreze & Sen (2002), although ME has grown in scope over time, we must still address concerns such as quality, equality, commercialization, and, above all, spiritual bankruptcy as areas of concern. There has been a great deal of recent discussion about the rapid expansion of private institutions worldwide. Education has risen to the status of an industry in global business. Because of its dynamic nature, the global competitive ME system faces several challenges in responding to sociological, technical, and economic developments in the local and international environment. While the World Trade Organization (WTO) plays a crucial role in opening up markets for trade in service sectors, the focus today is more on the significance and role of ME in a nation's social and economic development. However, the WTO lacks a clear strategy to enhance its education sector and compete with global giants. ME should aim to transform students into ethical, competent, and influential individuals who can drive their companies to greater heights while achieving business success.

In summary, ME in the new economy requires a unique perspective and vision that takes into account the evolving practices of executives in contemporary trades and intellectual management. Prompt attention must be given to address the emerging challenges and develop the necessary management skills to contribute to the country's growth and bring quality to modern businesses, ultimately positioning India as a global power.

According to Kakabadse & Vinnicombe (2018); and Prastacos et al. (2018), ME is among the most challenging matters facing organizations operating in the rapidly changing market. It is in desperate need of capable managers. During the 1990s, the Indian government liberalized the business education sector, resulting in a rapid increase in the number of b-schools offering full-time, part-time, executive, remote learning, or specialized courses. While ME has gained greater prominence due to the significant growth of management institutions, special attention should be given to the management practices through which business executives demonstrate the highest level of excellence and commitment in various global settings. To prepare these young individuals for self-sustainability and reform ME in India, the government, legislators, politicians, industrial companies, and social leaders have a significant duty to fulfill.

An array of issues must be addressed, some distinct from the others, but the majority are interconnected. These issues can be broadly classified, encompassing skills and knowledge associated with the overarching effect of behavioral characteristics. This is partly due to a scarcity of high-quality teaching staff combined with the enrollment of low-quality students in institutions that are inadequately prepared themselves. As a result of inadequate recognition and compensation, the teaching profession in India has struggled to attract top talent. This issue needs to be addressed with significant changes. Another critical issue is the gap between academia and industry. The program curriculum often lags behind and fails to align with the continuously changing requirements of the business world. Consequently, graduates from many institutions are not adequately prepared for the challenges they face at the start of their careers!

So, we are confronted with the dilemma of "number vs. quality," which we are currently witnessing. The quality of management programs varies significantly from one institution to the next. This is particularly apparent at lower-level institutions that lack well-trained teachers as well as the necessary infrastructure. Small- and medium-sized enterprises (SMEs) want to hire the most exemplary individuals; thus, they go for premier institutes like the former IIMs and ISB. Consequently, the existing system creates a "dual citizenship" of business executives, leading to unhappiness.

In addition to these significant structural and behavioral challenges, addressing the maintenance of program delivery quality is crucial. Other factors that continue to hinder the growth of this industry include a low rate of technology adoption and the use of antiquated pedagogical methods by management institutions, which are combined with legislative limitations.

9.3 PROBLEMS PREVAILING IN THE INDIAN MANAGEMENT EDUCATION (ME) SYSTEM: KEY POINTS

The following are the issues that exist in ME in Indian B-schools:

- The lack of quality curriculum and research in university departments is a critical flaw in Indian management education. There is a significant gap between the best and the rest of the institutions. In the top management school, the overall level of management education is declining.

- Another critical flaw in Indian management education appears to be a significant disconnect between the skills required for the job and what is taught in management colleges.
- The third flaw concerns instructional strategies that rely solely on the traditional and monotonous lecturing style.
- The academic audit is not available at management schools nationwide.
- Many Indian universities and management institutions deliver substandard management or business education.
- Academic administrators of management schools, colleges, and top universities are not chosen based on their overall merit and competence.
- Indian academic institutions are regarded as enclaves of excellence, as academics often have limited connections with societal structures or their global counterparts.
- Indian elite universities, business schools, and academic management institutions are highly centralized.
- The notion of participative management is not utilized in the decision-making process.
- Students are excluded from the decision-making process in the system of management education.
- In Indian business and management organizations, complex office processes are utilized.
- Administrative workers at business schools lack professional qualifications and training.
- The administration of Indian B-schools is based on strict rules.
- Many Apex Agencies' duties collide, including the AICTE, UGC, NAAC, NBA, NCTE, HMRD, NIEPA, and others.
- Lack of educational opportunities for diverse sectors of the population.
- Management schools do not have a standard cost structure (there is no common entrance exam for all management schools) and do not conduct or use student evaluations or feedback.
- Management schools do not make or utilize student evaluations or feedback.

9.4 GOOD PRACTICES ADOPTED IN MODERN-DAY B-SCHOOLS IN DEVELOPED COUNTRIES: KEY POINTS

The latest teaching approaches from advanced economies may assist in creating or improving a functional grasp of basic management science and technology. Hence, the characteristics of modern teaching techniques include:

1. **Learner-Centered:** One of the essential aspects of contemporary teaching techniques in management education is that they are learner-centered. They focus on learners when utilizing or applying them throughout classroom and laboratory courses. The instructor functions solely as a guide, and the learning process includes learners. Learners notably seem to dominate classroom discussions.
2. **Activity-based:** The instructor or guide plans the activity or assignment and encourages students to participate. As a result, it is a commission-based or activity-based system. Students are invited or required to participate in classroom engagement through these interactive exercises.
3. **Resource-based:** Teachers should gather and provide all necessary study materials to students in order to facilitate a thorough understanding of the subject. These materials can be obtained from the institution or any other accessible location. Additionally, learners themselves can also contribute by sharing study resources and materials.
4. **Interactive:** One feature that distinguishes current teaching methods is their highly participatory nature. Educators instruct learners to work in small groups or individually to accomplish the learning objectives and achieve desired outcomes. This approach facilitates learning from one another, helping students develop teamwork skills and a spirit of collaboration. Moreover, it benefits them as they navigate society.
5. **Integrative:** One of the most critical features of contemporary teaching approaches is integration. Teachers make it integrative by linking problems from one subject, such as drug usage, domestic abuse, security, contamination, distribution of food, and

criminality, to other concerns. A student may obtain knowledge of multiple subjects while studying one.

6. **Peer Collaboration:** Modern teaching techniques not only promote students by allowing them to express their ideas and take initiative, but also by monitoring their reactions, evaluating their research, and encouraging them to participate in classroom interactions. Additionally, these techniques select individuals based on their interests, needs, and feelings. Through classroom practices, learners collaborate and appreciate the efforts of their peers. Students' preferences are considered crucial in the curriculum, guiding them towards their goals and future careers.

9.5 CONCLUSION

The review was conducted to understand the significance of perspectives on difficulties and challenges in ME among college students and faculty members. This research demonstrates that educational institutions should support future growth in areas such as innovative teaching, off-the-job development, employability skills training, and addressing information and communication technology (ICT)-based issues, among others. Some MBA students will eventually become policymakers in their respective organizations. To promote economic growth and development in the country, it is necessary to allocate more time and resources to education. It also serves as a foundation for enhancing students' ability to manage commercial organizations and assume leadership positions in their respective countries.

Future ME requirements will differ from the past, yet today's business schools are too inflexible and conventional to adapt to new solutions. Many factors are working against change, and progress on one front will be insufficient to overcome the other. Furthermore, most business schools, rooted in a more prominent university and professional culture, have little influence over many of these roadblocks. The findings of this research shed light on the shortcomings of business and management schools. If management schools/universities remain vigilant, they will be able to quickly overcome their weaknesses and dangers by using their strengths and opportunities. It is also true that some of the issues mentioned above are difficult to put into practice. However, they remind us of the necessity to overcome difficulties and, as a result, assist in the formulation of clever strategies for obtaining recognition for Indian ME in the global context.

9.6 THE IMPENDING CHANGE AND THE PATH FORWARD

Due to the swift and dramatic changes that have influenced various aspects of the business environment, ME, particularly in India, is poised to undergo a revolutionary transition. The business world is seeking management talent with cutting-edge skills and expertise. This transformation is expected to significantly enhance the nation's design, content, and delivery of ME. The following are the aspects that we believe will be impacted by this impending transformation.

Fading boundaries, escalating competitive pressures: The COVID-19 pandemic has accelerated the pace of digitalization in education, which was already underway. With most programs now being conducted virtually, geographical borders have become irrelevant. Students can enroll in virtual courses offered by institutions in other countries. Many renowned international institutions have already established satellite offices in India, and their numbers are increasing. Additionally, several other institutions have partnered with Indian universities to offer joint degree programs. This competition is expected to intensify in the future, compelling Indian educational institutions to enhance and adapt their program design and delivery as quickly as possible.

Employers will increasingly demand that business school curricula match their business demands. Through more industry-focused specialist programs, recruiters will attempt to ensure that the b-school curriculum and employment needs are more closely aligned. As time goes on, super-specialized management credentials will be much more necessary! As a result, industry-specific specialty programs such as mutual funds and insurance, advanced analytics, health management, and other relevant subjects will expand in popularity. New programs and courses will arise, especially at the intersections of traditional functions of management (such as behavioral finance and analytic applications) and various industrial sectors (edge computing-based retail management, fintech, cloud-based logistics management). As a result, business schools will have to step up their efforts to meet the sector's demands.

The concept of "money worth of time" will guide the design of programs. There has already been a significant trend towards shorter-duration and modular programs that offer specialized competencies. This is expected to accelerate in the future as students desire to reduce their time learning new material. MBA programs lasting two-years will be phased out in favor of 15-month or one-year modules. Even shorter-duration programs that provide super-specialized certificates are expected to experience a surge in demand in the coming years. The challenge here will be to maintain the depth of program material and the high level of delivery while considering the time constraints.

Businesses will be compelled to embrace technology in new ways if they want to avoid perishing in the future. Transitioning to virtual modes of program delivery, re-skilling professors and staff to operate in a digital environment, and utilizing current pedagogical technologies such as simulation studies, big data analytics, VR technology, and blended learning, among others, will require a proactive strategy. There will be increased pressure on business schools to either improve or close their doors from recruiters and students. This pressure will be felt at all levels. The top international business schools will be forced to compare their programs against the top-tier colleges. As a result, the numerous intermediate and potentially lower-ranked schools will be pushed to enhance the efficiency and content of their facilities, expand their personnel pool, and improve interactions with graduates and employers.

Today, business school graduates are global citizens who seek career opportunities in various industries worldwide. If their degree-granting school has received globally recognized accreditation, their chances of being accepted increase. As a result, we anticipate that Indian business schools will increasingly pursue foreign certification from organizations such as the EMBA, AACSB, and Equis. This will enhance their worldwide exposure and improve their employability. India's regulatory bodies should support innovative experiments and advancements in ME, such as international collaborations, the freedom to offer joint degrees with foreign business schools, and the ability to select courses from diverse subjects across institutions.

The success of b-schools in India will depend on their ability to anticipate and adapt to changing requirements. Proactive organizations that analyze the shifting demands of the sector and collaborate with renowned universities to develop new courses will emerge victorious. This transition will benefit students, recruiters, educational institutions, and most importantly, our nation.

KEYWORDS

- **b-schools**
- **contemporary challenges**
- **economic growth**
- **educational institutions**
- **Indian management education system**
- **international collaborations**
- **management education**

REFERENCES

Azevedo, A., & Shane, M. J., (2019). A new training program in developing cultural intelligence can also improve innovative work behavior and resilience: A longitudinal pilot study of graduate students and professional employees. *The International Journal of Management Education, 17*(3), 100303.

Balamurugan, J., & Dharsini, L. P. (2017). Issues and challenges of management education in India: A global review. *International Journal of Applied Research, 3*(3), 162–165.

Bartlett, C. A., & Ghoshal, S., (1997). The myth of the generic manager: New personal competencies for new management roles. *California Management Review, 40*(1), 92–116.

Benson, S. N. K., & Ward, C. L., (2013). Teaching with technology: Using TPACK to understand teaching expertise in online higher education. *Journal of Educational Computing Research, 48*(2), 153–172.

Bhatnagar, N., & Ramachandran, K., (2021). *Management Education in India is Poised for a Major Transformation* | Business Standard News. Management education in India is poised for a major transformation. https://www.business-standard.com/article/management/management-education-in-india-is-poised-for-a-major-transformation-121011800044_1.html (accessed on 20 February 2024).

Canals, J., (2009). *Seeking a Greater Impact: New Challenges for Business Schools.*

Chakraborty, A., (2021). Foreign partnership in Indian higher education: Significance, challenges, and concerns. *Globalization, Societies and Education, 19*(5), 579–592.

Chauhan, S., Goyal, S., Bhardwaj, A. K., & Sergi, B. S., (2021). Examining continuance intention in business schools with digital classroom methods during COVID-19: A comparative study of India and Italy. *Behavior & Information Technology*, 1–24.

Child, J., (1996). *Management in China During the Age of Reform* (Vol. 23). Cambridge University Press.

Cohn, C., (2008). Mainstreaming gender in UN security policy: A path to political transformation? In: *Global Governance* (pp. 185–206). Palgrave Macmillan, London.

Collet, F., & Vives, L., (2013). From preeminence to prominence: The fall of US business schools and the rise of European and Asian business schools in the financial times global MBA rankings. *Academy of Management Learning & Education, 12*(4), 540–563.

Davenport, T. H., Prusak, L., & Wilson, H. J., (2003). *What's the Big Idea?: Creating and Capitalizing on the Best Management Thinking*. Harvard Business Press.

Day, C., Harris, A., & Hadfield, M., (2000). *Leading Schools in Times of Change*. McGraw-Hill Education (UK).

Dreze, J., & Sen, A., (2002). *India: Development and Participation*. Oxford University Press, USA.

Edumate Webinar: Management Education New Challenges/Opportunities – Edumate, (2021). Edumate. https://edumate.tv/edumate-webinar-management-education-new-challenges-opportunities/6610/ (accessed on 20 February 2024).

French, R., & Grey, C., (1996). *Rethinking Management Education*. Sage Publications, 2455 Teller Road, Thousand Oaks, CA 91320-2218 (hardcover: ISBN-0-8039-7782-4; paperback: ISBN-0-8039-7783-2, stock No. 4038; $26.95.

Fuller, C., (2002). The challenges and future opportunities of emergency management education: A student's perspective. *Australian Journal of Emergency Management, the, 17*(1), 24–26.

Hall, Jr. O. P., & Young, T. W., (2007). Management education benchmarking designing customized and flexible MBA programs. *Journal of College Teaching & Learning (TLC), 4*(3).

Harasim, L., (2006). A history of e-learning: Shift happened. In: *The International Handbook of Virtual Learning Environments* (pp. 59–94). Springer, Dordrecht.

Johannesson, J., & Palona, I., (2010). Asian megatrends and management education of overseas Chinese. *International Education Studies, 3*(1), 58–65.

Kakabadse, A., Bank, J., & Vinnicombe, S., (2018). *Working in Organizations*. Routledge.

Krishnamoorthy, B., (2017). *Environmental Management: Text And Cases*. PHI Learning Pvt. Ltd.

Lindell, M. K., Prater, C., & Perry, R. W., (2006). *Wiley Pathways Introduction to Emergency Management*. John Wiley & Sons.

Mahajan, R. N., & Sharma, V., (2012). Management education landscape in India: Issues, challenges, and strategies. *AIMA Journal of Management & Research, 6*(4/4), 1–12.

Mahajan, R., Agrawal, R., Sharma, V., & Nangia, V., (2014). Factors affecting quality of management education in India: An interpretive structural modelling approach. *International Journal of Educational Management*.

Mamun, M. A., & Mohamad, A. B., (2009). Management education for contemporary challenges: The role of business school. *European Journal of Scientific Research, 30*(4), 649–661.

Mitra, A., (2012). *Insights Into Inclusive Growth, Employment, and Wellbeing in India.* Springer Science & Business Media.
Muniapan, B., (2007). *Issues and Challenges of Management Education in Malaysia.*
Ottewill, R., & Macfarlane, B., (2003). Pedagogic challenges facing business and management educators: Assessing the evidence. *The International Journal of Management Education, 3*(3), 33–41.
Porter, L. W., & McKibbin, L. E., (1988). *Management Education and Development: Drift or Thrust Into the 21st Century?* McGraw-Hill Book Company, College Division, PO Box 400, Hightstown, NJ 08520.
Prastacos, G., Söderquist, K., Spanos, Y., & Van, W. L., (2002). An integrated framework for managing change in the new competitive landscape. *European Management Journal, 20*(1), 55–71.
Reddy, G. C., (2020). *A Study on the Academic Management of Fashion Education in India.* Doctoral dissertation, Maharaja Sayajirao University of Baroda (India).
Roddy, C., Amiet, D. L., Chung, J., Holt, C., Shaw, L., McKenzie, S., & Mundy, M. E., (2017). Applying best practice online learning, teaching, and support to intensive online environments: An integrative review. In: *Frontiers in Education* (Vol. 2, p. 59). Frontiers.
Rubens, A., Schoenfeld, G. A., Schaffer, B. S., & Leah, J. S., (2018). Self-awareness and leadership: Developing an individual strategic professional development plan in an MBA leadership course. *The International Journal of Management Education, 16*(1), 1–13.
Singh, S., & Jasial, S. S., (2021). Moderating effect of perceived trust on service quality–student satisfaction relationship: Evidence from Indian higher management education institutions. *Journal of Marketing for Higher Education, 31*(2), 280–304.
Thomas, J., & Mengel, T., (2008). Preparing project managers to deal with complexity–Advanced project management education. *International Journal of Project Management, 26*(3), 304–315.
Volery, T., & Lord, D., (2000). Critical success factors in online education. *International Journal of Educational Management.*

CHAPTER 10

Role of Applied Mathematics in Financial Management Education for Better Business Management

CHETNA

PG Department of Mathematics, M.M. Modi College, Patiala, Punjab, India

ABSTRACT

The rapid development of the Indian economy and the continuous expansion of market scale and business volume require the maturity of financial management theories and models not only at the enterprise level but also at the national level. The growing economic and financial management complexities require sophisticated and reliable solutions. Thus, it is necessary to continuously promote the development in financial management, which further requires the proficient use of applied mathematics. The relationship between applied mathematics and finance has attracted more and more attention from management researchers. The fundamental knowledge of applied mathematics has played an important role in the financial field by providing easy solutions to the problems that were previously difficult to address. The present chapter explores the relationship between applied mathematics and financial management. The chapter briefly analyzes the area of applied mathematics, then discusses the role of applied mathematics in modern financial management and points out the specific application of applied mathematics in modern financial management, aiding in better business management.

Interdisciplinary Approaches in Management Education.
Tanusree Chakraborty, Ramandeep Saini, Rajiv Khosla, Nandita Mishra, and Ashok Natarajan (Eds.)

10.1 INTRODUCTION

Whether it is in the subject of finance or in the research object, finance and applied mathematics are different, but this does not mean that there is no connection between the two. On the contrary, there is a close connection between finance and applied mathematics. Mathematical methods are ubiquitous and omnipresent in finance, such as partial differential equations, stochastic calculus, and stochastic processes. It can be seen that the research of finance needs to rely on applied mathematics to a certain extent, and the research and analysis of the relationship between the two is a relatively important area to explore and understand.

As an important branch of mathematics, applied mathematics encompasses a vast amount of theoretical knowledge and practical tools that have been extensively studied and applied in various social fields. It has particularly played a critical role in modern financial management. With the rapid development of the modern economy and the ever-changing market conditions, a thorough understanding of applied mathematical principles and their rational application in modern financial management activities can actively mitigate economic risks and provide essential guarantees for achieving greater economic benefits (Savard & Cavalcante, 2021).

Applied mathematics is a collective name for mathematical theories and methods that have a clear application purpose. As an essential branch of mathematics, it encompasses various fundamental and traditional mathematical theoretical knowledge. It also emphasizes the utilization of mathematical application methods in other fields and disciplines. For instance, differential equations, fuzzy mathematics, probability theory, and mathematical statistics can serve as theoretical application tools in social activities (Huston, 2010). With the increasing interdisciplinary collaboration, many applied mathematical theories and methods are frequently employed in economic research to aid in conceptual definition interpretations and to obtain logical research results, thereby effectively enhancing and advancing economic theories. Applied mathematics is commonly utilized in modern economic management activities, as modern economic management research and practice continue to progress.

The study is divided into various sections. The following sections come after the introduction:

- Quantitative relationship of financial activities;
- Promote financial entities to deepen reforms further;

- Collection and processing of financial management information;
- Financial and economic forecasts;
- Formulation of financial plans;
- Scientific management of the enterprise;
- Agroeconomics and applied mathematics;
- Financial management and game theory;
- Financial management and operations research;
- Finance management and probability theory;
- Financial management and calculus;
- Function model in financial theory;
- Limit theory in the financial economy;
- Derivatives in the financial economy;
- Differential equations in financial economics.

10.2 QUANTITATIVE RELATIONSHIP OF FINANCIAL ACTIVITIES

With the continuous deepening of the economic system, the main role of markets has become increasingly prominent, and the market situation has been changing very rapidly. If an enterprise wants to ensure stability in its production and development, it must effectively forecast the market economy situation and provide accurate information guidance for its economic activities. By utilizing quantitative mathematical analysis, companies can effectively adjust the quantitative relationship of economic activities to ensure that the company's production and supply align with market demands. Simultaneously, timely adjustment and control can prevent enterprises from blindly following trends and provide a safety guarantee to avoid financial losses for the companies (Sherraden & Ansong, 2016; Xiao & O'Neill, 2016).

Mathematical models can assist in the study of financial and economic activities. By utilizing scientific mathematical tools, we can continuously explore qualitative issues in the market economy, analyze and clarify quantitative relationships, and ultimately aid enterprises in better understanding and effectively utilizing modern economic laws (Wang, 2017). The theoretical conclusions derived from this research provide more specific practical applications for enterprises. The quantitative analysis of applied mathematics is an inevitable outcome of the development of financial science, and it also fulfills a practical need in the advancement of modern science and technology. The comprehensive understanding and rational utilization of applied mathematics methods in modern financial management are of great significance in optimizing the practical planning of enterprises.

10.3 PROMOTE FINANCIAL ENTITIES TO DEEPEN REFORMS FURTHER

Today, the Indian economy is in a transition phase, and it is necessary for business enterprises to properly implement appropriate reforms in the economic system to ensure healthy and sustainable development. The transformation of an enterprise from a product-oriented approach to an operational approach should align with its desired level of operations, development, and management. To overcome social competition, continuous self-optimization and reforms are essential (Agarwal & Mazumder, 2013). Applied mathematics can assist companies in understanding the current state of the social economy and the market, providing prediction, planning, coordination, and decision-making support for specific practical activities such as production, finance, sales, and operations in economic and financial activities. This enables companies to gradually deepen their business development reforms and achieve long-term growth (Lipman, 2003).

10.4 COLLECTION AND PROCESSING OF FINANCIAL MANAGEMENT INFORMATION

To effectively utilize applied mathematics in modern financial management, it is essential to have a comprehensive understanding of its various branches and specific application methods. Computer-oriented numerical methods, which are a part of applied mathematics, can assist companies in swiftly and accurately gathering and processing financial information. This facilitates prompt processing and precise information, thereby aiding business operations (Savard & Cavalcante, 2021). For instance, companies can employ search theory to gather financial management information, process the information using alternative set theory, and carry out information sorting and analysis through probability theory and mathematical statistics.

10.5 FINANCIAL AND ECONOMIC FORECASTS

Through the aforementioned collection and sorting of financial information, enterprises can obtain effective forecasts of their financial and economic activities, which in turn helps them make informed business decisions. For instance, companies can utilize time series methods and regression analysis to make scientific predictions based on management information and statistical data, and then conduct decision-making analysis by comparing the forecasted data with the actual market and company situation. This enables them to determine relatively accurate corporate decision-making and provide precise guidance for the enterprise's economic activities. By mastering the existing economic forecast system, enterprises can position themselves in the market competition (Johnson & Sherraden, 2007).

10.6 FORMULATION OF FINANCIAL PLANS

A small enterprise consists of multiple departments and organizations, while a large system like the national economy encompasses numerous departments and industries. Achieving a harmonious balance among these departments or enterprises is crucial. To accomplish effective macro-control, coordination, and balance, it is necessary to employ applied mathematical methods and develop corresponding research plans. For instance, the input-output method establishes the quantitative relationship between input and output using mathematical models, enabling analysis of macro- and microeconomic management. This analysis provides a reliable foundation for determining the proportional relationship between departments and production. By attaining a comprehensive understanding of balance, we can further devise optimal economic planning methods to facilitate the optimization of economic management (Cole et al., 2014).

10.7 SCIENTIFIC MANAGEMENT OF THE ENTERPRISE

Researchers have found that applied mathematics in modern enterprise economic management activities can provide enterprises with rich data and information supported by diversified mathematical modeling, ideas, and methods, thereby enhancing the scientific and rationality of enterprise decision-making (Xiao & O'Neill, 2016). Taking a tea company as an example, mathematical modeling methods are used to collect and organize relevant industry data to accurately analyze the problems existing in the development of the industry in order to deal with the external market competition situation. With the support of mathematical modeling methods, the production process can be optimized, the production cost of the enterprise can be reduced, and the economic benefit of the enterprise can be maximized. In the example of the tea business, the use of mathematical modeling methods can optimize the tea production process. Temperature, humidity, soil, and other factors directly affect the growth of tea. Mathematical models can be used to comprehensively collect and organize information about tea growth, and through modeling and adjustments, the growth environment of tea can be gradually optimized to increase tea production. Mathematical models can also be used in the marketing activities of tea products. Only through data analysis, the process of product marketing activities can be improved. Through the construction of mathematical models, the packaging design and image design of tea products can be analyzed and continuously adjusted and optimized to fully meet the needs of the public for value recognition (Zeng, 2015).

One of the key aspects of enterprise development lies in the vitality and vigor of the organization. The vitality of an enterprise is established through scientific management and organized planning. To achieve optimal economic management, it is necessary to utilize applied mathematics to formulate the best management plan and make reasonable arrangements for production, organization, and operational management. Probability theory, mathematical statistics, and related analysis ensure quality control, while storage theory aids in effective production cost management. Additionally, the application of business management theory in applied mathematics can enhance the marketing efforts of a company and comprehensively improve its business quality and management (Atkinson et al., 2007).

10.8 AGROECONOMICS AND APPLIED MATHEMATICS

As one of the important branches of mathematics, applied mathematics has a theoretical and practical foundation and can play a significant role in modern economic management. Whether it is national economic management or business management, it is necessary to fully implement theoretical and practical research on applied mathematics, refine the application tools, and better implement them in economic management activities. Applied mathematics can also deepen the depth of economic research through its rigorous theoretical analysis models and quantitative analysis methods. Additionally, it can effectively improve the accuracy, authenticity, and meticulousness of economic research conclusions (Zeng, 2015). The area of financial economics and applied mathematics are inseparable in the following ways:

1. **Combinatorics:** Combinatorial mathematics, also known as discrete mathematics, focuses on processing various discrete data using algorithms. In the current era of rapid computer technology development, combinatorial mathematics plays a crucial role in enhancing computer efficiency when dealing with discrete objects. For instance, consider a scenario where a salesman needs to visit N regions to sell agricultural products in agricultural economics. How can we minimize the distance traveled to ensure all regions are covered? Assuming N has a value of 20, even with a computer capable of processing hundreds of millions of calculations per second, it would take at least 100 years for a sub-speed computer to solve this problem (Zeng, 2015). Combinatorics can optimize the computer's algorithm for solving such problems, significantly reducing the computation time, and further advancing research in this field.

2. **Mathematical Statistics:** This primarily studies effective methods for collecting, organizing, and analyzing data that are influenced by random factors. It aims to provide scientific and reasonable speculations and judgments, which serve as important references for specific decision-making actions. In the field of agricultural economy, the growth and development of crops in experiments often vary due to the influence of ecological environment and various random factors. This, in turn, affects the reliability and authenticity of experimental results. Mathematical statistics can address this by considering specific experimental conditions, selecting the most scientific and reasonable experimental design and sampling techniques, and utilizing various tools and methods such as parameter estimation, hypothesis testing, variance analysis, and regression analysis. By doing so, it can achieve a higher

degree of accuracy in estimation and judgment, thereby further promoting the development of the agricultural economy (Savard & Cavalcante, 2021).

3. **Fuzzy Mathematics:** It is also a critical component of applied mathematics. As the name suggests, fuzzy mathematics focuses on studying and processing fuzzy phenomena. Fuzzy cluster analysis and fuzzy comprehensive evaluation are common methods used in fuzzy mathematics. In the field of agricultural economy, there are many uncertainties in climatic conditions, disaster detection, variety selection, land resource classification, and other aspects. By applying fuzzy mathematics in applied mathematics, various problems can be scientifically solved. For instance, factors such as the number of green leaves, seedling height, rhizome length, and thickness often directly impact the growth and appearance of flax. By utilizing pattern recognition in fuzzy mathematics, the quality of a flax plant can be accurately determined based on these factors. Another example is the extraction of characteristics such as ear date, effective ear number, plant height, 100-kernel weight, and main ear kernel number through pattern recognition, allowing for accurate judgment without knowledge of specific wheat varieties. It is evident that there is a close connection between applied mathematics and economics, particularly in agricultural economics (Savard & Cavalcante, 2021). With the aid of applied mathematics, the use of rigorous and standardized data collation, analysis, and inference methods can effectively address various agricultural economics issues and expedite the establishment and development of modern agricultural science (Paul & Elder, 2001).

10.9 FINANCIAL MANAGEMENT AND GAME THEORY

Game theory is primarily a mathematical method and theory used to study phenomena characterized by competition or conflict. It examines the interaction between formulated incentive structures (Zhang, 2012). The "smart pig game" serves as a classic example of game theory in finance. The main scenario involves two pigs, one large and one small, in a pigpen. The pig trough is located on one side of the pen, while the pigs are positioned at the opposite end of the trough. There is a button on the other side of the pen that controls the supply of pig food. Pressing this button increases the amount of food in the trough by 10 units. However, each pig consumes two units of food energy while making its way to the trough. If the larger pig reaches the trough first, the ratio of food between the large and small pig is 9:1. If the smaller pig reaches the trough first, the ratio becomes 6:4. If two pigs simultaneously press the button to control the pig's food, the ratio of big pigs to little pigs in relation to the pig's food is 7:3. When these two pigs possess a certain level of wisdom, the outcome is that the smaller pig will undoubtedly choose to wait. By utilizing the payout matrix in game theory, it can be calculated that if the smaller pig opts to wait, the worst-case scenario results in a return of 0, while the best-case scenario yields a return of +4. Conversely, if the smaller pig chooses to act, the best-case scenario results in a profit of 1, but there is still a risk of a –1 loss. Therefore, if the pig is wise, it will unquestionably choose to wait. This is known as the famous "smart pig game" (Gao & Yao, 2018). This classic case and conclusion can be applied to today's finance. For instance, a small company can choose to remain patient and wait when necessary, allowing a larger company to take the lead in market development. Inaction during this time can pave the way for future achievements. By choosing to wait, small companies can save significant unnecessary costs in research and observation, enabling the company's development and management to reach new heights.

10.10 FINANCIAL MANAGEMENT AND OPERATIONS RESEARCH

Operations research was practically born in the 1930s. It can be considered an emerging discipline. When managers make decisions, operations research can provide an important scientific foundation. The goal of operations research is to achieve effective management and accurate decision-making. It is an extension of applied mathematics, utilizing mathematical models, statistics, and algorithms to find optimal or near-optimal solutions to complex problems. In normal circumstances, operations research is primarily used to analyze and optimize more complex real-world problems. In finance, many errors are caused by an excessive selection of constraints in the mathematical model. Optimization is a crucial aspect of operations research, with the majority of problems focusing on studying optimization methods. The optimization method can be determined carefully for the optimal constraints in the mathematical model (Abidin et al., 2010).

10.11 FINANCE MANAGEMENT AND PROBABILITY THEORY

Based on the current situation, finance research has gradually evolved from static research to dynamic research, with a shift in the direction of exploration. Simultaneously, a deep understanding of random phenomena has become increasingly important. Probability theory primarily focuses on studying random phenomena as a mathematical subject, while mathematical statistics is mainly the direct application of probability theory. In finance, econometrics serves as the main embodiment for applying mathematical statistics. Econometrics is a discipline that relies on actual data and utilizes mathematical statistics to create corresponding economic models. It combines real-world economic phenomena to establish certain relationships or equations, which are then further determined using actual data. Additionally, economic loss estimation is another application of probability theory in finance. As an important branch of finance, the development of insurance concepts is based on probability statistics. Probability statistics knowledge has been widely employed in parameter estimation and insurance property loss assessment (Savard, 2018).

10.12 FINANCIAL MANAGEMENT AND CALCULUS

There is a significant relationship between finance and game theory, operations research, mathematical statistics, and probability theory. Additionally, there is a strong connection between finance and calculus. Financial activities involve the interaction of various quantities, encompassing multiple elements in this communication process, with 'function' being one of them. When examining the function in isolation, it appears as a simple mathematical problem. However, when placed in a real-life context, the performance of the function in financial activities becomes highly complex. The various quantities are intricately intertwined, resulting in a complicated relationship. Sorting out this relationship in a short period of time or expressing it quickly is not an easy task. On the other hand, the relationship among derivatives, calculus, and actual variables is relatively straightforward and determinable. As we know, the calculus equation comprises essential elements such as the derivative, the unknown function, and the independent variable. It is precisely because the calculus equation incorporates these two important elements, function, and derivative, that we can observe the extensive utility of calculus in financial activities (Savard, 2017; Feudel & Biehler, 2021).

The function of a calculus equation may exist in two or more forms in actual economic problems. This differs somewhat from the theoretical knowledge in finance. When practical financial problems involve calculus equations with two or more unknown functions, one of the functions can be treated as a constant variable. This simplifies the calculus equation, as it can be processed using the knowledge of a single variable. In this process, the bias theory in derivatives is applied. This demonstrates the inseparable relationship between finance and calculus. When dealing with practical economic problems and calculus, we also utilize basic theoretical knowledge such as differentiation and total integration (Feudel & Biehler, 2021).

10.13 FUNCTION MODEL IN FINANCIAL THEORY

In mathematical theory, functions serve as fundamental building blocks of knowledge. When studying mathematical methods in the field of financial economics, it is often necessary to establish specific functional relationships and analyze economic phenomena by examining the interplay between these functions. This allows us to solve problems effectively. For instance, when investigating the relationship between supply and demand within a market system, relevant theories in economic mathematics can be employed to establish a functional relationship between supply and demand through a functional model. By calculating the results at a given stage, we can clearly define the dynamics between supply and demand. Functions elucidate the connection between market supply and demand. The supply function can be considered as a dependent variable. As the price of a product increases, the supply will also increase, while the demand will decrease. Conversely, the dependent variable can also be a demand function. In a market economy, the economic law dictates that value determines price, which in turn influences sales volume. Through this functional relationship, we can identify the equilibrium point of market supply and demand, providing a foundation for decision-making and fostering the prosperity and development of the financial economy (Savard & Cavalcante, 2021).

10.14 LIMIT THEORY IN THE FINANCIAL ECONOMY

The core of mathematics is the theory of limits, which serves as the foundation for many mathematical concepts. The theory of limits is frequently employed in the analysis of various aspects of an economy. Its significant role lies in the development and reduction of entities. For instance, the cell limit theory plays a crucial role in the development and reduction of various phenomena, such as the growth and decline of species, population growth and decline, and cell growth and division. Limit theories are commonly utilized in financial investment management and other fields for the economic analysis. For example, limit theories can be applied to calculate bank deposit annuities and compound interest (Savard, 2017).

10.15 DERIVATIVES IN THE FINANCIAL ECONOMY

A theory often used in mathematics is the theory of derivatives, which has a close relationship with economics. The concept of marginal is used to establish the relationship between derivatives and economics, allowing for a more comprehensive study of economic phenomena by replacing constants with variables. Derivatives are commonly employed in economics, particularly in functions such as marginal revenue, marginal cost, and marginal demand, which are frequently used when analyzing economic factors.

The derivative reflects the change in the independent variable and analyzes the change in the dependent variable by varying the independent variable to obtain the rate of function change. When studying the cost function, the marginal cost can be calculated by fixing the output of the product; that is, the cost of producing an additional unit of the same product. In this case, the marginal cost can be compared to the average cost to determine the change in output for this product. If the marginal cost is lower than the average cost, it means that the output of this commodity has increased. If the marginal cost is higher than the average cost, the output is reduced. Another application of derivatives is in elasticity research. Elasticity research can study the changes in functions and the relationship between demand and commodity prices.

The corresponding price value is obtained through elasticity, and the price ratio of the increase of the commodity is higher than the reduction ratio of the demand. As a result, the enterprise will gain more benefits due to the increase in price. However, if the commodity price exceeds this value and the commodity increase ratio is lower than the demand reduction ratio, then an increase in price will lead to a reduction in the company's revenue. Economic optimization is an important part of economic analysis, which can also be analyzed through derivative theory. Problems such as optimal resource allocation, optimal income, and maximum profit can be solved by determining the extreme value and maximum value of the derivative (Feudel & Biehler, 2021).

10.16 DIFFERENTIAL EQUATIONS IN FINANCIAL ECONOMICS

A key component of economic mathematics is the differential equation, and its effective application can solve some problems in economics. In financial and economic analysis, the relationship between one quantity and another often arises, which can be analyzed and resolved through functional relations. When encountering very complex functional relations, it is necessary to analyze and solve them through differential equations (Savard, 2017).

Differential equations include unknown functions, differentials, independent variables, etc., and are one of the functional relationships. Analysts cannot use derivatives to accurately reflect quantitative relationships when analyzing complex financial and economic issues. In this way, they can be applied to embody differential equations. However, due to the great difficulty of differential equations and the complexity of the content, analysts should pay special attention to avoid missing information when applying them. By doing so, the role of differential equations in the financial economy will be fully utilized and will contribute to economic research and analysis (Savard & Cavalcante, 2021).

10.17 CONCLUSION

In conclusion, we can say that in finance, the role of applied mathematics is immense. Applied mathematics provides indispensable research tools and models that play an essential role in promoting the development of financial theories. Applied mathematics is simply a collection of mathematical theories and models with a clear application purpose. It is a vital branch of mathematics, containing basic and traditional mathematical theoretical knowledge. However, more is needed to study how to apply it, including differential equations, fuzzy mathematics, numerical methods, probability theory, mathematical statistics, and many other branches of mathematics knowledge into other categories. Therefore, we can also think of applied mathematics as the development and extension of traditional mathematics, especially in economic research, which often requires a large amount of professional mathematics knowledge. By analyzing and interpreting various concepts and definitions with the help of applied mathematics, we can obtain more intuitive research results under rigorous logical thinking, improving and promoting existing financial and economic theories. The chapter briefly explains the relationship of mathematics with finance from an application perspective.

KEYWORDS

- **applied mathematics**
- **financial economics**
- **financial theory**
- **game theory**
- **mathematical theories**
- **modern financial management**
- **quantitative relationship**

REFERENCES

Abidin, R. A. Z., Alwi, F., & Jaafar, N. F. M., (2010). A case study on mathematics anxiety for mathematics of finance course. *US-China Education Review, 7*(11), 1–11.

Agarwal, S., & Mazumder, B., (2013). Cognitive abilities and household financial decision making. *American Economic Journal: Applied Economics, 5*(1), 193–207.

Atkinson, A., McKay, S., Collard, S., & Kempson, E., (2007). Levels of financial capability in the UK. *Public Money & Management, 27*(1), 29–39.

Cole, S. A., Paulson, A., & Shastry, G. K., (2014). Smart money? The effect of education on financial outcomes. *Review of Financial Studies, 27*(7), 2022–2051.

Feudel, F., & Biehler, R., (2021). Students' understanding of the derivative concept in the context of mathematics for economics. *J. Math. Didakt, 42*, 273–305.https://doi.org/10.1007/s13138-020–00174-z

Gao, X., & Yao, Y., (2018). The application of economic game theory – taking the actual phenomenon as an example. *European Journal of Research and Reflection in Management Sciences* (Vol. 6, No. 1). Progressive Academic Publishing, UK.

Huston, S. J., (2010). Measuring financial literacy. *The Journal of Consumer Affairs, 44*(2), 296–316.

Johnson, E., & Sherraden, M. S., (2007). From financial literacy to financial capability among youth. *Journal of Sociology & Social Welfare, XXXIV*(3), 119–145.

Lipman, M., (2003). *Thinking in Education* (2nd edn.). Cambridge University Press.

Paul, R., & Elder, L., (2001). *Critical Thinking: Tools for Taking Charge of Your Learning and Your Life*. Prentice Hall.

Savard, A., & Cavalcante, A., (2021). Financial education and mathematics education: A cross-cutting analysis of the epistemological intersection of financial numeracy. *Mathematics Education in the Digital Era, 15*. Springer, Cham. https://doi.org/10.1007/978–3-030-73588-3_1.

Savard, A., (2017). Making decisions in a complex world: Teaching how to navigate using mathematics. In: Michelsen, C., Beckmann, A., Freiman, V., & Jankvist, U. T., (eds.), *Mathematics as a Bridge Between the Disciplines; Proceedings of Mathematics and its Connections to the Arts and Science International Symposium (MACAS 2017)* (1–14). Copenhagen, Denmark.

Savard, A., (2018). Teaching probability and learning financial concepts: How to empower elementary school students in citizenship. In: Cooter, K. S., & Lucey, T., (eds.), *Financial Literacy for Children and Youth* (2nd edn., pp. 137–152). Peter Lang.

Sherraden, M. S., & Ansong, D., (2016). Financial literacy to financial capability: Building financial stability and security. In: Aprea, C., Breuer, K., Davies, P., et al., (eds.), *International Handbook of Financial Literacy* (pp. 83–96). New York: Springer.

Wang, K., (2017). Analysis of the relationship between applied mathematics and finance. *Curriculum Education Research*, (30), 257.

Xiao, J. J., & O'Neill, B., (2016). Consumer financial education and financial capability. *International Journal of Consumer Studies, 40*, 712–721. https://doi.org/10.1111/ijcs.12285

Zeng, J., (2015). Analysis of the application of economic mathematics in financial economic analysis. *Journal of Jilin Radio and Television University*, (4), 45, 46.

Zhang, W. Y., (2012). *Game Theory and Information Economics*. Shanghai: Shanghai People's Publishing House.

CHAPTER 11

Can Interdisciplinary Management Education Lead to Achieving Sustainable Development Goals? A Conceptual Framework

AMIT KUMAR,[1] AMEET SAO,[1] GAURAV ASHESH,[2] and ERDEM UTKU EKE[3]

[1]*RICS School of Built Environment, Amity University, Noida, Uttar Pradesh, India*

[2]*School of Business, AURO University, Surat, Gujarat, India*

[3]*Assistant Professor, Nigde Omer Halisdemir University, Department of FEAS, Nigde, Turkey*

ABSTRACT

The 1990s saw a surge in the demand for management education (ME) due to strong economic growth. As a result, a large number of public and private institutions across the country began offering ME programs. However, the proliferation of these institutions has its own structural flaws. Additionally, there are numerous obstacles within the sector, which can be categorized as diverse skills and knowledge that have a significant impact on behavioral elements. This is the result of a combination of a scarcity of high-quality faculty members and the enrollment of low-caliber students at certain under-equipped institutions.

The fundamentals of an SDG-driven responsible management education (SDG-RME) include the elements of purpose, values, method, research,

Interdisciplinary Approaches in Management Education.
Tanusree Chakraborty, Ramandeep Saini, Rajiv Khosla, Nandita Mishra, and Ashok Natarajan (Eds.)

partnership, and dialogue. ME can be seen as a catalyst for sustainable development, where the four pillars of sustainability – human, social, economic, and environmental – play a crucial role. An SDG-RME instills a sense of responsibility in young minds, motivating them to appreciate these sustainability concepts and work towards improving their respective organizations and society.

The takeaways from this discussion can be understood through the lens of an SDG-RME Ecosystem, considering both the driving and restraining forces. Ultimately, quality education serves as the foundation for building strong nations and robust economies. ME is not an isolated activity. Its overarching purpose is not just to earn profits, but to work towards achieving sustainability as well. The key to success lies in ensuring that all the parties with which a business is interacting stand to gain – customers, management, society, government, regulatory bodies, investors, among others.

11.1 INTRODUCTION

The 1990s witnessed a significant increase in the demand for management education (ME) due to robust economic growth. Consequently, numerous public and private institutions across the country initiated ME programs. The MBA degree has emerged as a highly sought-after qualification, often considered a solution for various business challenges (BS, 2021). However, the proliferation of these institutions has its own structural flaws, leading to inherent problems. These obstacles can be categorized as diverse skills and knowledge that have a significant impact on behavioral elements. This is a result of a combination of factors, including a scarcity of high-quality faculty members and the enrollment of low-caliber students in certain under-equipped institutions. Due to the undervaluation and underpayment of teaching in many of these institutions, they struggle to attract highly knowledgeable workers. Addressing this issue requires fundamental improvements (BS, 2021).

11.1.1 HIGHLIGHTS OF THE CHAPTER

The fundamentals of an SDG-driven responsible management education (SDG-RME) are presented as follows:

1. **Purpose:** Management education (ME) should work towards conceiving and facilitating a society that is sustainable in the long run.

2. **Values:** Management education should instill values designed around the ethos of global social responsibility (GSR) into the students so that they evolve as the torchbearers of GSR in their respective domains (Thoresen, 2010, 2012).
3. **Method:** An SDG-RME should aim to develop educational frameworks, resources, and methods in order to create an environment that promotes effective learning.
4. **Research:** An SDG-RME should promote relevant research with respect to redefining the role of an organization in achieving social and environmental sustainability.
5. **Partnership:** Management education has the pivotal responsibility of interacting with the corporate sector in order to gain a better understanding of the challenges faced by potential employers. Brainstorming on remedial measures may help preemptively address these issues.
6. **Dialog:** Opportune dialogue should be held in the areas of concern to the organizations and companies so that appropriate solutions may be devised.

11.2 MANAGEMENT EDUCATION (ME) AS A CATALYST FOR SUSTAINABLE DEVELOPMENT

No development may be perceived as tenable unless it is sustainable (Kravets et al., 2018; Miles, 2018). There are four basic pillars of sustainability:

1. **Human Sustainability:** This aims to improve the condition of human beings in society. Diverse businesses affect society in various ways. SDG-RME plays a key role in promoting human sustainability by developing the skills and competencies of individuals, enabling the growth of both the organization and society.
2. **Social Sustainability:** This aims to preserve social capital and strengthen the basic framework of society. It helps conceive a better society for future generations to thrive in.
3. **Economic Sustainability:** No business can survive in the long run without earning a certain profit. Hence, economic sustainability is important to ensure the company's existence and growth.

4. **Environmental Sustainability:** This aims to conserve soil, water, and other resources as they are key to achieving meaningful progress in the long run.

   SDG-RME instills a sense of responsibility in young minds so that they feel motivated to appreciate the above-mentioned constructs of sustainability and work towards the betterment of the respective organization as well as society (Flogaitis, 2011).

11.3 MANAGEMENT EDUCATION (ME) FOR ERADICATION OF IMPOVERISHMENT

While education is the backbone of a progressive society, economic growth in a country can only be achieved through good education. The lack of jobs in the public sector has placed the entire burden of employment generation on the private sector. Countries have made significant investments in improving the education system and human capital to achieve economic development (Ozturk, 2001). When it comes to managing businesses, ME is considered extremely crucial due to its significant influence in shaping human capital according to future needs. The quality of managers working in the corporate sector plays a vital role in determining the success or failure of respective business entities. Good ME contributes to the establishment of new businesses and the improved performance of existing ones, thereby creating meaningful employment opportunities. This phenomenon increases the earning power in an economy and boosts the disposable income of the masses. The economic development resulting from these actions helps in eradicating poverty. A company established in a new region or emerging sector acts as a booster for the economy. For instance, the Tata Motors' plant in Gujarat had a positive impact on the local economy, leading to faster growth as ancillary industries also flourished in the same region, generating substantial employment opportunities, and significantly reducing poverty levels.

11.4 MANAGEMENT EDUCATION (ME) AND 'ZERO HUNGER'

Despite a consistent decrease in the years leading up to 2014, the number of people affected by hunger has shown a gradual increase. As of 2020, the pandemic and human conflicts have resulted in 265 million individuals suffering from hunger (UNDP, 2018). The elimination of hunger continues to be one of the most pressing issues in developing countries like India, where the high population further exacerbates the problem.

Providing access to ME to individuals would be immensely beneficial in the future. It would help develop a new generation that is industry-ready, capable of contributing to the workplace, and generating suitable employment opportunities for others as well. Providing appropriate training to the entire population is an immensely ambitious task. However, it would be instrumental in generating the much-needed surplus in a given economy. Moreover, it can encourage individuals to foster a performance-oriented environment.

11.5 MANAGEMENT EDUCATION (ME) FOR A HOLISTIC WELLNESS

Having a good lifestyle and prosperity is crucial for sustainability. The world is currently facing an ongoing crisis in the form of human conflict following the COVID-19 pandemic, which has further weakened the socio-economic framework in various countries. The healthcare infrastructure of many nations has been exposed to such a crisis, revealing previously unnoticed loopholes. While significant advancements have been made, such as improved life expectancy and reduced mortality rates among children and women, more concerted efforts are now needed as a wide and evolving range of diseases pose a fresh threat to humanity.

ME can promote a more socially responsible approach in organizations by contributing appropriately towards these causes. The corporate social responsibility (CSR) norms facilitate the well-being of a given society by ensuring the availability of the much-required funds and other resources. ME can foster a sense of belongingness and commitment in diverse managerial teams regarding their duty to give something back to society as well.

11.6 MANAGEMENT EDUCATION (ME) AND 'QUALITY EDUCATION'

Education goes a long way in eradicating poverty. Over the previous decade, significant advancements have been made towards improving the reach of education to rural and semi-urban areas, especially for girls.

Various initiatives have been taken by the Central and State Governments, in addition to different NGOs, to encourage and facilitate education for all.

However, the quality of education has suffered due to various geopolitical crises from time to time. The pandemic triggered numerous challenges for educational ecosystems worldwide. It was particularly intense for developing countries, where knowledge workers had insufficient access to Internet-enabled educational platforms. The rapid growth of institutions further exacerbated the problem of educational quality. Fortunately, institutions have gradually but consistently realized that quality is the key to survival and growth. The focus on the quality of education and research has paved the way for creating a highly intellectually stimulating environment at educational institutions. They have recognized the importance of both national and international accreditations, as well as the level of monitoring and mentoring involved. This has established a well-defined process of teaching and learning at the institutions.

11.7 MANAGEMENT EDUCATION (ME) AND 'GENDER EQUALITY'

Gender equality is important for prosperity and sustainability across regions and cultures. In the past, fewer girls attended school and even fewer women assumed leadership roles. However, the scenario has now changed with the help of suitable legal provisions, administrative and institutional interventions, as well as socio-political evolution.

Nevertheless, many challenges still persist in diverse systems, which keep women as the underrepresented gender at various levels and domains. While the crime against women remains alarmingly high, the COVID-19 episode has further worsened the imbalances for women – from well-being to security and social assurance. The prolonged online working schedules have also resulted in a drastic increase in workload for women. Those women who work in the unorganized sector have been exposed to immense pressure since the pandemic triggered their chances of falling into abject penury.

ME has enhanced awareness and consciousness towards women's representation in various departments and sectors in general, and leadership roles in particular. The corporate sector is also favorably responding to this change. This is evident in the increasing number of female students being admitted to educational institutions and moving towards gainful employment. The reservation for women has further facilitated holistic efforts to bridge this inequality in society. Despite various institutional and governmental schemes and special measures that have addressed this issue to a significant extent, gender equality continues to pose challenges across regions and cultures.

11.8 MANAGEMENT EDUCATION (ME) AND 'CLEAN WATER AND SANITATION'

Irrespective of the substantial efforts made towards enhancing the accessibility to potable water and better sanitation, a large majority of the population continues to have limited access to these essential services. The COVID-19 pandemic has increased awareness of hand hygiene and highlighted the importance of reducing crowding in public places to maintain both basic and medical hygiene. While hand washing is one of the best ways to reduce the chances of infection, it requires the overall availability of clean water as well.

ME has successfully explored and emphasized the importance of efficiently and optimally managing such services and resources to maximize their reach. Many management professionals are working in this area to provide quality services and ensure their availability to the masses.

11.9 MANAGEMENT EDUCATION (ME) AND 'AFFORDABLE AND CLEAN ENERGY'

The world is working towards ensuring the production and distribution of energy to as much of the populace as possible. The efforts are directed towards the concept of energy being green as well as more accessible across regions, irrespective of less fortunate countries and economies. Energy efficiency is now the new mandate for organizations and systems across diverse parts of the world.

The notion of energy efficiency requires countries to track progress in the following domains:

- Energy access;
- Energy efficiency;

- Renewable energy.

The availability of energy is important for the fight against pandemics – current and potential ones. It is also crucial for ensuring seamless healthcare facilities and clean water. Energy acts as an enabler for uninterrupted IT services during difficult times like COVID-19, which required people to maintain social distancing and access telemedicine.

ME has encouraged a thorough revamping of this sector by raising awareness about the inefficient modes of power generation. The traditional modes of power generation are not just costly but also unsustainable in the long term. Research studies conducted in this domain have contributed significantly to developing the understanding that renewable energy is strongly recommended. Now, companies and organizations have realized the importance of newer and more sustainable forms of energy as various models of clean energy ecosystems are being explored.

11.10 MANAGEMENT EDUCATION (ME) AND ENABLED WORKPLACES AND PROGRESSIVE ECONOMIES

An augmented model of workplaces and economies is the backbone for sustainable development, helping to improve the socio-economic standards of the populace across the world (Quoquab & Mohammad, 2017). Better employment opportunities are generated with economic growth, which in turn ensures economically secure countries. The wage gap among different nations around the world also reduces significantly in accordance with higher levels of global economic growth.

As per the statistics, a substantial proportion of the global population earns a living on USD 2 a day, compared to USD 1.8 in the case of India. It should be noted that obtaining any job does not guarantee a transition above the poverty line. A conducive external environment is necessary to provide better opportunities for the masses. Adequate global investments contribute to improved consumption habits and patterns, as well as the availability of healthy investment avenues in a specific economy. Therefore, prioritizing the creation of new jobs through prudent economic policies and development plans can generate decent and robust work ecosystems. Additionally, labor laws must be fully observed across all regions. This strategy may reduce disparities in workplaces and societies while enhancing the growth rate in the given economy (Washington, 2016).

It may be noted that "preserving the environment is a crucial measure to support sustainable economic growth. The natural environment contributes directly, by providing resources and raw materials such as water, timber, and minerals that are required as inputs for the production of goods and services; and indirectly, through services provided by ecosystems including carbon sequestration, water purification, managing flood risks, and nutrient cycling" (UNDP, 2018).

Any nation's routine activities and economic growth plans are affected by natural disasters and human conflicts. The resultant volatility in growth throws many households into poverty. Hence, environmental concerns must be borne in mind in order to achieve economic development. Creating and maintaining a good ecosystem would go a long way in creating a tenable socio-economic development.

11.11 MANAGEMENT EDUCATION (ME) AND 'INDUSTRY, INNOVATION, AND INFRASTRUCTURE'

Industrialization is a driving force for development and requires a robust infrastructure to ensure sustainability. It is important to note that location-based infrastructure can lead to significant economies of scale, promoting inclusive growth. India is actively attracting foreign direct investment (FDI) and forging strategic partnerships to foster inclusive industrial development. The country is investing its resources to establish a sustainable and economically sound framework.

The advent of digital technologies in the fintech sector has brought about a profound transformation in access to financial services. Mobile money, crypto assets, market lending, and insurance tech are among the rapidly evolving phenomena in this domain. This innovation is seen as the democratization of technology, driving change across various sectors. It has facilitated advancements such as AI, machine learning, and IoT, which have proven to be a boon for the startup ecosystem. Entrepreneurs are capitalizing on these opportunities, creating progressive employment prospects for skilled workers. Investments in renewable energy, eco-friendly solutions, and the built environment will further generate employment opportunities and yield additional benefits.

Furthermore, nuclear technology, despite its potential hazards in the event of a disaster, boosts industrial productivity and energy security. It is used to clean contaminants from industrial wastewater and enhance air quality, thereby promoting sustainable development. The industry has recognized the importance of sustainability and is moving towards green buildings. More LEED and GRIHA Certifications have been witnessed, depicting the interest of individuals and organizations to pursue the green cause (Smith, 2015).

By contrast, a poorly designed system of innovation can exacerbate biodiversity loss and lead to economic decline. ME plays a crucial role in understanding technical expertise in this era. The skills acquired through this education are aimed at ensuring optimal resource utilization to create and sustain a better world order. Techno-managerial capabilities enable individuals to comprehend the potential and opportunities for innovation within a specific system. They facilitate proper analysis and problem-solving, which in turn expands the market for developing better infrastructure. These concepts and practices encourage new establishments to embrace and promote innovation. Additionally, ME helps identify solutions to complex financial and infrastructural challenges while building capacity in line with the sustainable development goals (SDGs). Professional consensus establishes the strategic vision and direction that can generate value and contribute to the nation's progress.

11.12 MANAGEMENT EDUCATION (ME) AND 'REDUCED INEQUALITIES'

Developing countries like India often face two well-known structural problems: high youth unemployment and high inequality. Growing global inequality has become a major barrier to the growth and transformation into a developed nation. Inequality increases when the population of a country does not have equitable access to education, healthcare facilities, and legal protection by the state. Inequality can be classified into two broad categories:

1. **Horizontal Inequality:** It refers to inequality arising from groups that exist in diverse societies, such as religion, among others.
2. **Vertical Inequality:** It refers to inequality arising out of income groups, in addition to other factors.

   Education has a significant impact on the human resources of a country. It fosters the principles of social justice in individuals and contributes to the reduction of inequality. The role of ME in this field is subject to thoughtful deliberation due to the diverse range of opinions on this matter. However, current research and thought-provoking ideas are increasingly focused on creating a comprehensive and effective knowledge base. It is a domain where concepts and applied knowledge are closely intertwined. Taking into account the competencies of each individual, domain knowledge should be applied in practical scenarios to ensure that employment opportunities occur within a real-life context. Education is a priority sector for the government, and the budgetary expenditure on education has been progressively increased with the objective of reducing inequality and unemployment in society.

11.13 MANAGEMENT EDUCATION (ME) AND 'SUSTAINABLE CITIES AND COMMUNITIES'

Urbanization is undoubtedly the main driver of economic growth. However, rapid urbanization has led to an increase in slum-dwelling and overburdened infrastructure. Additionally, the significant population growth has made the issue of insufficient housing a major concern and a highly debated topic in national development. Statistics indicate that the hybrid working model resulting from the pandemic has greatly increased the demand for housing. The low interest rates have further contributed to this trend. Initiatives such as the Housing for All Mission (2015), PMAY-2015, and SWAMIH-2019 have brought significant investments at the national level (Kumar & Shukla, 2022).

India is making efforts to ensure that 80% of housing development activities fall under the affordable segment (Nallathiga et al., 2018). Additionally, the establishment of a sustainable transport system is considered crucial as it accounts for approximately 25% of energy consumption related to greenhouse gases. This emphasizes the importance of an intelligent public transport system that addresses the major mobility needs of the population, thereby creating a positive impact. This SDG also emphasizes the preservation of cultural and natural heritage due to its significance.

India has established a strong partnership with the least developed countries (LDC) through its robust support and assistance mechanisms targeted towards this specific region. However, it is important to note that all policies and regulations related to growth and development are constantly evolving, requiring progressive social and structural changes. Effective managerial skills are crucial in leading this process of change management, ensuring the efficient utilization of regional resources. This approach promotes enhancements in quality and an increase in per capita income.

Skilled experts should assess the bottlenecks in the growth process and contribute to the potential for economic development. Developing appropriate strategies and consistently monitoring quality could help reduce regional disparities. Given the scale and seriousness of the emerging threats, ME offers collective support in staying relevant by analyzing the current situation to achieve a better future.

11.14 MANAGEMENT EDUCATION (ME) AND 'ETHICAL PRODUCTION AND CONSUMPTION (EPC)'

EPC involves "doing more and better with less," enhancing welfare output and living standards by minimizing waste and optimizing the use of resources (Liarakou, 2019). A few figures that explain the current scenario in India are shared in this context:

- "Third largest greenhouse gas emitter, which translates to 6.9% of global emissions;"
- Only 19.9% of the urban India's waste is being processed;
- 54.6% of the power generation is sourced from coal;
- While the country is home to 18% of the world's population, it only possesses 4% of global water resources (SDG Tracker, 2018).

The United Nations proposes that "Responsible consumption and production is about promoting sustainable infrastructure, resource and energy efficiency, by providing access to basic services, green and decent jobs, and a better quality of life for all" (UNDP, 2018). The adoption of the above-mentioned SDG aims to achieve growth targets by reducing environmental concerns, economic costs, social costs, poverty, and thus enhancing economic competitiveness (Miller, 1995).

The implementation of these goals requires a well-planned approach taken by all stakeholders, working together from production to eventual consumption (UNESCO, 2005). It also requires educating people, through various mediums, about adopting sustainability as a lifestyle. ME plays a major indirect role in achieving these goals through various initiatives and measures, as follows (Liarakou, 2019):

- Include themes, topics, modules, courses, and degrees about education for responsible consumption in the established curriculum;
- Encourage research in education for responsible consumption-related areas;
- To enhance cooperation between professionals from diverse disciplines in order to develop integrated approaches to education for sustainable consumption;
- Reward creative, critical, and innovative thinking related to education for sustainable consumption;
- Ensure that education for responsible consumption respects the importance of indigenous knowledge and recognizes alternative lifestyles;
- Provide opportunities for the practical application of theoretical study through social involvement and community service.

11.15 MANAGEMENT EDUCATION (ME) AND 'CLIMATE ACTION'

Management education (ME) can help develop a better understanding and implementation strategy for climate action. This knowledge strengthens resilience against climate-related hazards prevalent in the built environment. ME can improve awareness and institutional capacity through the study of real-life case studies and brainstorming activities.

Early adaptation to change reduces response time for management trainees to become productive in their respective industries. Management studies can highlight the need for effective planning in managing climate-related changes. This agile methodology and corresponding mechanism can protect and promote the interests of all stakeholders.

Continuous systemic checks, the ability for quick adaptation, and real-time control are key features that education can provide at the primary, secondary, and tertiary stages. Specialized training programs focused on climate action and synergies can generate plausible and meaningful change in the system.

Students play a key role in learning evolving technology and its interface with management. It is essential to develop the nation's sustainable development strategies and framework in accordance with the industry's growth imperatives, without damaging the climate and compromising profitability. This gap can be filled with a better understanding of economic and social concerns (UNDP, 2018; AACSB, 2015).

11.16 MANAGEMENT EDUCATION (ME) AND 'LIFE BELOW WATER'

The research capabilities of management institutions provide a sense of direction to the industry in accordance with their requirement of advancing sustainability. SDG-RME helps develop a thorough understanding of the oceans and seas, their opportunity cost, the small island states, and sustainable tourism. It requires a two-fold perspective from both sustainability as well as business profitability aspects. A thorough understanding of all kinds of pollution in the sea, debris, and nutrient pollution, along with a strategy to curb the pollution at the industry level, is needed in order to manage and protect the marine and coastal ecosystems. ME can help formulate frameworks and bridge the gap in these areas. Such a focus will minimize and address the impact of ocean acidification, overfishing, and illegal fishing.

By setting achievable goals, a strategy can be formulated for the conservation of the marine and coastal areas. This sustainable model could benefit the smaller countries engaged in tourism, fisheries, and associated activities. Data warehousing and knowledge management systems facilitate the transfer of scientific knowledge and guide interventions to improve the health of the marine ecosystem. Adequate resources and more accessible markets may be provided without compromising the delicate balance of such resources. ME can train the upcoming workforce to fill this vital gap (AACSB, 2015; UNDP, 2018).

11.17 MANAGEMENT EDUCATION (ME) AND 'LIFE ON LAND'

The care for the environment and the profitability concern for business should not be perceived as watertight compartments. Businesses cannot exist as an isolated entity devoid of this fundamental concern for the environment. It is important to understand the comprehensive relevance of any business in terms of its risks and opportunities. Any business endeavor and land-based production activity must consider the importance of forests and their resources. Managers must grasp the significance of biodiversity, protected areas, and the ecosystems they support.

A prudent balance between conservation and extraction is crucial in this context. Managers can promote and implement best practices for conserving all types of forests while also increasing forest cover. A robust framework for sustainable forest management involves a holistic understanding of the soil, its nature, and geology.

Managers need to ensure that the benefits arising from better resource utilization are shared across both domestic and foreign markets. Management studies can help integrate intrinsic values with the planning and development process to maintain a judicious balance. Working closely with the industry, they can mobilize and enhance financial resources to build and sustain other resources while providing adequate incentives for all stakeholders.

Management of land and its varied but responsible use can also promote the development of local communities by building their capacity to pursue sustainable livelihood opportunities. This is one of the prime domains where management studies can make a meaningful impact (AACSB, 2015; UNDP, 2018).

11.18 MANAGEMENT EDUCATION (ME) AND 'PEACE, JUSTICE, AND STRONG INSTITUTIONS'

Management education (ME) should perceive these SDGs as an opportunity to align the learning outcomes and curricula with research priorities. This mindset will facilitate a shift in the desired direction, enhancing the industry's relevance to society. SDGs have a long-term framework, presenting challenges for implementation in areas such as peace, justice, and institutions worldwide. A well-designed ME program can contribute to the development of robust systems that are free from corruption.

Furthermore, the cultivation of critical thinking and effective communication can lead to better stakeholder responses. It is crucial to promote inclusive and participatory decision-making at all levels within the entire system. Management studies become even more relevant when national and international institutions engage in meaningful collaboration and cooperation in areas that impact the social environment.

The ensuing process helps to build capacity, prevent violence, and combat crime. This approach promotes peace and enforces the relevant laws and regulations to ensure sustainable development and growth. Prudent policies and proactive actions by leading institutions generate trust among various stakeholders. Management studies play a primordially important role in imparting ethics, driving a sense of belongingness and responsibility, and promoting transparency in the system. These managers then evolve as trustworthy and reliable knowledge workers with respect to their thoughts and actions (UNDP, 2018; AACSB, 2015).

11.19 MANAGEMENT EDUCATION (ME) AND 'PARTNERSHIPS FOR THE GOALS'

The SDGs communicate a well-deliberated and actionable message to all businesses worldwide, with the mandate to collaborate with other disciplines and develop a wide range of opportunities. In this context, it is important to establish new and innovative partnerships with various institutions and industries across the globe. This can be achieved by creating platforms for different stakeholders to come together for meaningful discussions and radical innovations.

Considering divergent views can help in resolving imminent issues more effectively. Emphasizing interdisciplinary work leads to deeper collaboration, which can address problems and challenges faced by many businesses. Aligning academic goals with sustainable goals contributes to the betterment of the community. This approach enhances capacity building, ensures commitment to growth, promotes egalitarian trade, and facilitates sustainable development strategies.

Management studies can further strengthen resource mobilization and improve revenue generation capacity. Coordinated policies are essential in providing assistance for developing strategies and frameworks to achieve long-term goals.

ME can also help in promoting the development and transfer of technologies that support the SDGs and are beneficial for all stakeholders. These initiatives would foster a universal, rule-based, and open environment, ultimately enhancing the economic growth of the country. An SDG-RME (sustainable development goals-related management education) could emerge as a crucial factor in facilitating public and private partnerships through collaborative experiences, comprehensive implementation strategies, and continuous monitoring. Management schools need to design a set of initiatives to establish measurable goals and facilitate milestones, thereby supporting the capacity-building framework in the required areas (UNDP, 2018; AACSB, 2015).

11.20 KEY TAKEAWAYS

The takeaways of this discussion can be illustrated in Figure 11.1, which depicts the continuum of an SDG-RME Ecosystem in the context of driving forces and restraining forces:

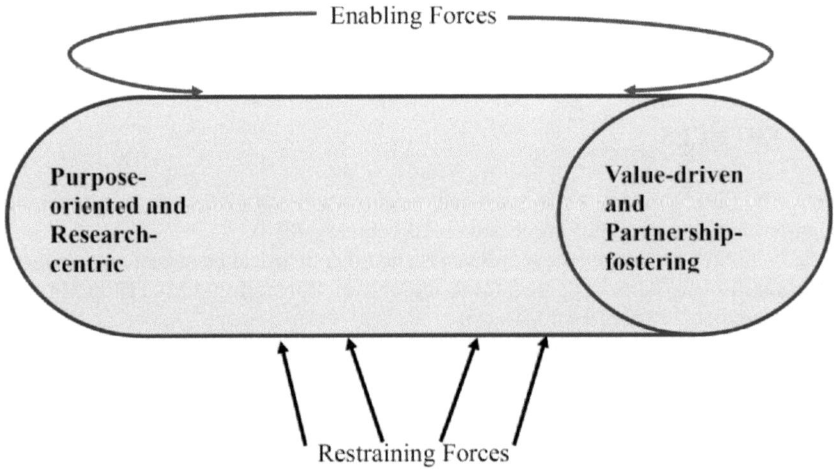

FIGURE 11.1 The continuum of an SDG-RME ecosystem.

Source: Authors' compilation.

11.21 CONCLUSION

Quality education serves as the foundation for strong nations and thriving economies. ME is not an isolated endeavor; its overarching purpose extends beyond profit-making to encompass sustainability. Success lies in ensuring mutual benefit for all stakeholders involved in business interactions, including customers, management, society, government, regulatory bodies, and investors. By adhering to these guiding principles, businesses can achieve efficiency and pave the way for future generations to follow.

KEYWORDS

- **climate action**
- **ethical production and consumption**
- **gender equality**
- **management education**
- **quality education**
- **sustainable development**
- **zero hunger**

REFERENCES

AACSB, (2015). https://www.aacsb.edu/insights/articles/2022/05/integrated-approaches-to-teaching-sustainability (accessed on 20 February 2024).

BS, (2021). https://www.business-standard.com/article/management/management-education-in-india-is-poised-for-a-major-transformation-121011800044_1.html (accessed on 20 February 2024).

Flogaitis, E., (2011). *Education for the Environment and Sustainability*. Pedio, Athens. ISBN 978-9-6095-5226-4.

Kravets, O., Maclaran, P., Miles, S., & Venkatesh, A., (2018). *The SAGE Handbook of Consumer Culture*. Sage, London. ISBN 978-1-4739-2951-7.

Kumar, & Shukla, (2022). Affordable housing and urban poor in India. *Social Change, 52*(1). 58–75.

Liarakou, G., (2019). Education for responsible consumption and sustainable development. *Encyclopaedia of Sustainability in Higher Education*. Springer, Cham.

Miles, S., (2018). The emergence of contemporary consumer culture. *The SAGE Handbook of Consumer Culture* (pp. 11–26). Sage. ISBN 978-1-4739-2951-7

Miller, D., (1995). *Acknowledging Consumption*. Routledge, London. https://doi.org/10.4324/9780203975398. ISBN 978-1-134-84312-1.

Nallathiga, R., Lele, S., & Prasad, V., (2018). *Affordable Housing Development in India: Current Models and Their Replication, 19*, 43–57.

Ozturk, I., (2001). *The role of education in economic development: A theoretical perspective. Journal of Rural Development and Administration, XXXIII*(1), 39–47.

Quoquab, F., & Mohammad, J., (2017). Managing sustainable consumption: Is it a problem or panacea? Sustainable economic development. *World Sustainability, Series* (pp. 115–125). Springer International Publishing Switzerland. ISBN 978-3-319-45081-0.

SDG Tracker, (2018). *Sustainable Development Goals Tracker*. https://sdg-tracker.org/sustainable-consumption-production (accessed on 20 February 2024).

Smith, R. M., (2015). Green building in India: A comparative and spatial analysis of the LEED-India and GRIHA rating systems. *Asian Geographer, 32*(2). doi.10.1080/10225706.2015.1020065.

Thoresen, V. W., (2010). *HERE and NOW! Education for Sustainable Consumption-Recommendations and Guidelines*. UNEP, Paris.

Thoresen, V. W., (2012). Developing value-based, holistic education for sustainable living. In: *Proceedings: Global Research Forum on Sustainable Consumption and Production Workshop*. Rio de Janiero, Brazil.

UNAIR Repository, (2018). *Sustainable Development Goals*. https://www.undp.org/sustainable-development-goals (accessed on 20 February 2024).

UNDP, (2018). *Sustainable Development Goals*. https://www.undp.org/sustainable-development-goals (accessed on 20 February 2024).

UNESCO, (2005). *International Implementation Scheme*. Report by the Director-General on the United Nations Decade of Education for Sustainable Development. https://unesdoc.unesco.org/ark:/48223/pf0000140372 (accessed on 20 February 2024).

Washington, H., (2016). Introduction. Why the growth economy is broken. *The Future Beyond Growth; Towards a Steady State Economy* (pp. 1–14). Routledge, London. ISBN 978-1-138-95302-4.

CHAPTER 12

Interdisciplinary Management Education and Sustainable Development: Key Insights

GEETANJALI SINGH,[1] INDU CHAUHAN,[2] and YUSUF HASSAN[3]

[1]*Assistant Professor, P.G. Department of Economics, Mehr Chand Mahajan DAV College for Women, Chandigarh, India*

[2]*PhD Scholar, Department of Economics, Kurukshetra University, Kurukshetra, Haryana, India*

[3]*Assistant Professor–HRM, University of Birmingham, Birmingham, United Kingdom*

ABSTRACT

Interdisciplinary is one of the most popular keywords in 21st-century academia. While it is not a new notion, its significance has become more apparent in recent years. Interdisciplinary approaches offer numerous advantages in the workplace, including collaboration, communication, research, flexibility, interpersonal skills, and other multidisciplinary abilities that are highly sought after today. Education plays a key role in society and has a significant impact on a country's overall socioeconomic growth. The sustainable development goals (SDGs) set time-bound targets for various sectors crucial to development, such as health, education, employment, sanitation, energy, infrastructure, and the environment. From a global perspective, India's achievement of these goals is crucial. The importance of inclusion and high-quality education for all is reaffirmed.

Interdisciplinary Approaches in Management Education.
Tanusree Chakraborty, Ramandeep Saini, Rajiv Khosla, Nandita Mishra, and Ashok Natarajan (Eds.)

Providing quality education for everyone emphasizes the idea that sustainable development is a proven and powerful tool in achieving education for all. This chapter aims to provide insights into the evolution of interdisciplinary concepts in management education (ME) over time. The SDGs mandate that governments ensure inclusive and equitable quality education for all children and adults, as well as lifelong learning opportunities. This, combined with providing access, aims to promote equity, inclusiveness, and sustainable education. Therefore, the objective of this chapter is to study the SDGs in the Indian context and examine how they are integrated with ME. It examines the status of goal attainment in education in India at various levels, including primary, secondary, and tertiary, and seeks to identify some of the major challenges faced by the education sector in providing quality education and addressing accessibility issues.

12.1 INTRODUCTION

The advantages of an interdisciplinary approach are plentiful. Students who possess the ability to understand and connect various knowledge and skills are better equipped for rewarding careers and a more fulfilling life. Management education (ME) and interdisciplinarity were relatively unknown in the academic and commercial worlds until the 1950s. Management was not considered a separate profession. The imaginary and, in fact, detrimental concept of despecialization suffocated new ideas about preparing young people for careers in business. This quickly changed following the introduction of American ME principles into the Marshall Aid framework (Van Baalen). One of the goals of the Sarva Shiksha Abhiyan (SSA) is to achieve complete primary education for children attending school and beyond. As a result, the Twelfth Five Year Plan (2012–2017) prioritized the expansion, improvement of quality, and provision of equal educational opportunities for all sections of society. With an average age of 29 years, India has an advantage over the rest of the globe when it comes to the graying population. The demographic dividend can be tapped by improving health, education, and skill development.

This study examines the visible patterns in Indian education at the primary, intermediate, and postsecondary levels. Additionally, it discusses several significant issues in delivering quality education and challenges in accessibility in ME. Over the past five years, India has implemented a number of SDG (sustainable development goals) initiatives. Both the Central and State governments have used the SDGs as guiding principles for development activities at various levels. Some of the flagship programs include Swachh Bharat Abhiyan, Ayushman Bharat, and Poshan Abhiyan, among others. These programs are designed to align with the SDG targets (NITI Aayog, 2021). The SDGs have aimed to address inter-sectoral tensions while integrating the mandates of different actors by bringing together economic, social, and environmental issues. Consequently, policy coherence requires the development of awareness and expertise among government officials at all levels to establish and manage effective implementation methods. Through a series of 25 national and sub-national consultations, as well as specific engagement with various states and Union Territories (UTs) on SDGs and their implementation strategies, NITI Aayog, in collaboration with Union ministries and the UN system, has taken forward sensitization and awareness building from the start. Many state governments, in collaboration with district administrations and civil society, have devised and implemented a variety of public awareness and capacity building initiatives for the public and officials, respectively, based on local situations. Collaborations like these support partnerships as a key strategy for moving the SDGs forward at all levels of government and administration. Civil society has been included as a vital partner to attain the targets outlined in the 2030 Agenda for sustainable development (UNESCO, 2015).

A comprehensive examination of the history of science reveals that the discipline is inherently contingent. Consequently, various alternative metrics have been employed to depict the development of numerous scientific fields, encompassing different domains, areas, research frameworks, scientific specialties, and other paradigms. Some scholars advocate for studying disciplines from a diachronic perspective, given the historical nature of the discipline being contingent, and the challenge of establishing consistent criteria for demarcating scientific fields (Bechtel). This perspective recognizes that social and cognitive factors interact to shape identity, domain, and context. Throughout history, disciplines have established arbitrary boundaries that fluctuate over time (Gieryn, 1983), varying across countries and institutions. Therefore, the institutionalization of science into disciplines should be understood in terms of spatial and intellectual continuities and transitions, rather than a linear progression and organization of scientific knowledge into distinct fields (Wagner and Wittrock). Academic disciplines are deeply embedded in the institutional context of national higher education (HE) systems. This chapter discusses the value of interdisciplinary studies in ME. New scientific ideas have emerged from interdisciplinary research efforts, such as operations research, management games, instructional design, game theory, dynamic programming, and mathematical information theory, which are fundamentally formalistic and abstract.

Sustainability addresses economic, social, environmental, and governance concerns in all nations, leaving no area unattended. The SDGs play a crucial role in ME as they have the power to influence students and shape tomorrow's business leaders and society through responsible and sustainable business practices. These objectives in ME help identify the potential of future companies and provide business school professors with a diverse range of topics to enhance the relevance and impact of research. SDGs in ME can contribute to the development and preparation of current and future business practitioners and company leaders. Business schools can also support policymakers and the general public in formulating, evaluating, and identifying practical and feasible policies. By producing graduates who are knowledgeable about sustainability and fostering students' intellectual and change making capabilities, business schools are increasingly making an impact on global development. This may have an impact on students' ability to engage in responsible and sustainable business activities, shaping the corporate leaders and society of tomorrow.

The benefits of an interdisciplinary approach are numerous. Students who can comprehend and link a wide range of knowledge and abilities are on their way to more fulfilling lives and employment opportunities. This study examines the trends in Indian education at the primary, intermediate, and postsecondary levels. Additionally, it discusses several important issues in delivering quality education and addressing accessibility in ME.

Sustainability addresses economic, social, environmental, and governance concerns of all nations, ensuring that no one is left behind. The SDGs play a crucial role in ME as they have the power to impact students and shape tomorrow's business leaders and society through responsible and sustainable business practices.

Business schools are playing an increasingly influential role in global development by producing graduates who are knowledgeable about sustainability and capable of driving change. This has a direct impact on students' ability to engage in responsible and sustainable business activities, shaping the future corporate leaders and society.

12.1.1 AN OVERVIEW

Interdisciplinary education in management can be described as follows. An interdisciplinary approach is required in business studies to address the challenges. Interdisciplinary relies on the convergence of various realities. Interdisciplinary business studies will showcase advanced techniques of creativity and the ability to generate new ideas as thinking and technology progress.

In December 2018, the first edition of the SDGs India Index was released, which included 62 indicators from 39 targets across 13 SDGs. The index was later updated to a second edition in December 2019, encompassing all the aforementioned aspects. This version featured more indicators, including 68 that were fully aligned with the National Indicator Framework (NIF), 20 that were refinements from the previous version, and 12 from other official government sources. The current edition, marking the third, represents an improvement over the 2019–2020 edition with a broader coverage of targets.

The Index provides progress reports on the social, economic, and environmental sectors of the country and its constituent States/UTs in their pursuit of achieving the SDGs. The index has been designed to be accessible to everyone, including policymakers, civil society, businesses, and the general public.

The India Index for SDGs includes the following goals:

- To rank all the Indian states as well as the Union Territories (UTs) in order of performance.
- Encourage healthy competition among Indian states and UTs in their pursuit of the global goals.
- To assist states and UTs in determining priority regions.
- To enable states and UTs to learn from the best practices of their counterparts.
- To identify data gaps in the states' and UTs's statistical systems and the areas in which more robust and frequent data is needed (NITI Aayog, 2021).
- Understanding that providing universal access to high-quality education reinforces the view that one of the most powerful and time-tested drivers of long-term development is education. By 2030, this goal aims to ensure that all boys and girls have completed their basic education, i.e., elementary and secondary school, and that they can benefit from affordable vocational skill development training.

- This also addresses the issue of gender disparity and income inequality, while also paving the way for high-quality education, with a special emphasis on higher education (HE).
- It also emphasizes lifelong learning opportunities in order to attain substantial adult literacy and numeracy, as well as the creation and development of education facilities that are sensitive to the age of the learner (child), any disability, and gender.
- Economic progress may not be sustained until a growing proportion of the population is self-assured and capable of standing on their own due to mastery of information and skills.
- Education improves people's access to economic opportunities.

12.2 CURRENT APPROACHES IN SDGS: QUALITY EDUCATION IN INDIA

Education is the most important component of the SDGs for long-term development. All of the objectives are very ambitious and comprehensive in nature, with the aim of being relevant to all people around the world to ensure that "no one is left behind." Within a specific time frame, these goals aim to eradicate poverty, eliminate hunger, improve health and education standards, achieve gender equality, provide clean water, proper sanitation, and sustainable energy, and combat climate change. As a signatory to the United Nations SDGs, India has committed to providing all children with inclusive and high-quality education by the year 2030.

Because of its interconnections with the other SDGs, ME is a multi-dimensional element for achieving most of the goals (SDGs). Assessing and monitoring educational quality is challenging. However, it should encompass various areas of education, including individuals' development in terms of personality, cognitive talents, and non-cognitive talents, the development of various skills, knowledge creation and transmission, and the development of values and attitudes. For a long time, the focus has been on infrastructure facilities such as school buildings, boundary wall construction, classrooms, drinking water, and toilets, and metrics have been generated based on these features. Among all the parameters discussed above, teachers are considered the most significant input in the learning mechanism, and therefore, emphasis should be placed on timely recruitment of teachers and avoiding prolonged vacancies.

12.2.1 THE SUSTAINABLE DEVELOPMENT GOALS (SDGS) ON EDUCATION (SDG 4)

These goals aim to ensure that all children have access to quality education up to the secondary level, and that all adults have access to technical and academic education at the university level. Additionally, these goals encompass providing suitable education to people with disabilities, tribal and native populations. They also include special provisions for countries that are not fully developed, as well as smaller nations, especially those situated on small islands (United Nations General Assembly, 2015, p. 17).

If India is to meet the targets set by Goal No. 4 of the SDGs within the 15-year time frame envisioned by the UN, it must seize the window of opportunity to provide accessible, quality education. This will enable Indian children to attain minimum proficiency levels in reading and mathematics, ensuring that India has a skilled and productive labor force to support its aspirations of becoming a global leader (Niti Aayog).

To achieve these targets, specific indicators have been established for each key component of SDG 4. These indicators include enrollment ratios at the elementary and secondary levels, dropout ratios for children aged six to 13, as well as at the secondary level, minimum proficiency levels based on nationally defined learning outcomes, gross enrollment ratios in HE, gender parity index (GPI), and the percentage of disabled children attending school. Additionally, indicators related to the proportion of trained teachers at various levels and target pupil-teacher ratios have been identified. Together, these indicators form the India Index, and Annual Reports provide a summary of the status for each indicator, serving as the basis for the succeeding sections.

12.2.2 STATUS OF INDIA'S PROGRESS IN FULFILLING GOAL 4 OF THE SDGS

1. **Enrolment Ratio in Elementary Education:** The Adjusted Net Enrolment Ratio at the Elementary level for Classes 1 to 8 was 87.26%. Only one union territory and three states were able to achieve the target of 100% enrollment at the elementary stage. These states are Karnataka, Manipur, Tripura, and Delhi. Among the states, Nagaland had the lowest enrollment ratio of 67.38%,

while among the UTs, Jammu and Kashmir and Ladakh had the best ratio of 67.88%.

2. **Average Annual Dropout Rate:** The average annual dropout rate at the secondary level in India was 17.87%. The best performing state was Himachal Pradesh with a 7.81% dropout rate, and Chandigarh was the best performing UT with a 4.52% dropout rate. Arunachal Pradesh had the highest percentage of dropouts at the secondary level at 35.98%, followed by Assam at 31.47%, and Andaman and Nicobar Islands with 22.22%.
3. **Enrolment Ratio in Higher Secondary:** The gross enrolment ratio (GER) at the higher secondary level was found to be 50.14%. Himachal Pradesh was the best-performing state again with a recorded GER of 81.79%, while Bihar had the lowest GER of 26.39%. Among the UTs, Chandigarh had the highest gross enrollment ratio of 83.43%.
4. **Student Proficiency in Class 8:** Overall, 71.9% of students achieved at least the minimum proficiency level in Language and Mathematics, based on nationally defined learning outcomes at the end of grade VIII. The state of Rajasthan had the highest performance, with 88.10% of students achieving the minimum proficiency level. Among the UTs, Chandigarh performed the best with 81.6%. However, only 60% of Class 8 students achieved the minimum proficiency level in Arunachal Pradesh, Jammu and Kashmir, and Ladakh, respectively.
5. **Enrolment Ratio in Higher Education (HE):** For the 2018–2019 academic year, the percentage of students between the ages of 18 and 23 enrolled in HE was 26.3%. Sikkim was among the states and Chandigarh among the UTs that had the highest enrollment ratio at 53.9% and 50.6%, respectively. Among the states, Bihar had the lowest enrollment at 13.6%.
6. **Education Level Among Persons with Disability:** 19.3% of individuals aged 15 and above with a disability had completed at least their secondary education. No state or union territory had achieved the target of 100%. The state with the highest performance was Goa at 32.4%, and the union territory with the highest performance was Delhi with 41%. On the other hand, only

10.3% of individuals with disabilities in Arunachal Pradesh had completed secondary education.

7. **Gender Parity in Higher Education (HE):** An all-India GPI value of 1 is reported according to the AISHE report 2018–2019, implying parity between females and males (18–23 years) in HE in the country. Around 15 states and all the UTs have achieved gender parity in HE. The GPI was lowest in Bihar at 0.79.
8. **Literacy Levels:** The literacy level among individuals aged 15 years and above at all India level stood at 74.6%. Mizoram, at 98.5%, and Lakshadweep, at 95.7%, are the closest to achieving the target of 100% literacy. In Bihar, only 64.7% of individuals aged 15 years and above were literate.
9. **Infrastructure in Schools:** In 2018–2019, in terms of access to basic infrastructure, 84.76% of schools had complied with the target of providing electricity and drinking water. Goa, at 99.97%, and Gujarat, at 99.95%, are the closest to achieving the target of 100%. Only 28.4% of schools in Meghalaya have access to basic infrastructure. Among the UTs, Chandigarh, Dadra and Nagar Haveli, Daman and Diu, Delhi, and Lakshadweep had achieved the target.
10. **Proportion of Trained Teachers:** Approximately 83% of teachers in secondary schools are trained. The percentage of trained teachers among the states is highest in Kerala and Puducherry (94.53%), and (92.57%) among the UTs. Assam has less than 30% trained teachers at the secondary level.
11. **Pupil-Teacher Ratio:** The all-India pupil-teacher ratio at the secondary level stands at 21. The target is to have at least one teacher for every 30 students in all schools. Himachal Pradesh has the best pupil-teacher ratio in the country with one teacher for every 9 students at the secondary level. Six states, namely Bihar, Gujarat, Jharkhand, Madhya Pradesh, Uttar Pradesh, and West Bengal, are yet to achieve the target.

    It should be noted that an Act of government, such as the right of children to free and compulsory education (RTE), may have provisions to provide inclusive elementary education for all. This

Act specifically intends to give equality in opportunities for all sections of society. Thus, the Act intends to cover gender, Scheduled Castes, Scheduled Tribes, migrants, minorities, out-of-school children, and children with special needs. However, for RTE to work, strong government intervention is needed.

12.3 EDUCATION AND SUSTAINABLE DEVELOPMENT GOALS (SDGS): INTER-SECTORAL LINKAGES

The concept of sustainable development emerged from a growing concern that human society had a negative impact on the environment and needed to address it. In 1987, the Brundtland Commission (officially known as the World Commission on Environment and Development) defined sustainable development as "development that meets the needs of the present without compromising the ability of future generations to meet their own needs" (Brundtland, 1987). The 2030 Agenda for Sustainable Development, along with the adoption of the SDGs, aims to integrate and balance the economic, social, and environmental aspects as the three pillars of sustainable development.

The expansion of school infrastructure, the reduction of gender disparities in literacy, and the overall literacy rate have all contributed to progress in education. Education plays a crucial role in empowering women and girls for the sake of building a sustainable future. Education is interconnected with all other SDGs and has several specific targets related to poverty, hunger, nutrition, universal health coverage, gender equality, decent work, and climate action. Therefore, education is the most important multidimensional component for achieving most of the SDGs.

Literacy and educational attainment are fundamental indicators of a society's level of development. Literacy is often associated with key aspects of modern civilization, such as urbanization, industrialization, communication, and trade. Literacy is a vital element in an individual's overall development as it enables them to better understand and respond to their social, political, and cultural environment. Higher levels of literacy and education lead to increased awareness and improvements in economic and social conditions.

Thus, we argue that in order to achieve a more sustainable world, a good education is a necessary component. Therefore, Education for

Sustainable Development (ESD) facilitates the development of the factors required to build an environmentally friendly, socially fair, and economically viable world. These factors include understanding sustainability, embracing sustainability values, acquiring the skills and knowledge to progress towards sustainability, as well as engaging in activities that promote sustainability. The social and human rights dimension of ESD is expanded. Hence, ESD is a dynamic idea that encompasses a new vision of education aimed at empowering individuals to take responsibility for ensuring a sustainable future. The overarching goal is to empower citizens to act for positive environmental and social change, which requires a collaborative and action-oriented strategy (UNESCO, 2002) (Table 12.1).

TABLE 12.1 School Education: Elementary

Parameter	2010–2011	2019–2020
Enrolment (crores)	19.3	26.45
Number of schools (lakhs)	13.62	42.3
Dropout rate (%)	6.5	0.11
GER * (primary)	118.62	97.8
(Upper primary)	81.15	89.7

Source: NITI Aayog (2021).

However, the gross enrollment rate (GER) in primary education fell from 118.62% in 2010–2011 to 97.8% in 2019–2020, whereas the GER in upper primary education climbed from 81.15% to 89.7% during the same time period. The percentage of students that drop out of school has decreased from 6.5% in 2010–2011 to 0.11% in 2019–2020 (Table 12.2).

TABLE 12.2 Adult Literacy Rate (%; Age 15 Years and Above)

Year	Total	Male	Female	Gender Gap
2001	61	73.4	47.8	25.6
2020	77.7	84.7	70.3	14.4

Source: GOI, MHRD, Educational Statistics at a Glance (2021).

From 61% in 2001, the general literacy rate grew to 77.7% in 2020. Male literacy rates increased by 11.3% points, rising from 73.4% in 2001 to 84.7% in 2020, while female literacy rates increased by 22.5%

points, rising from 47.8% in 2001 to 70.3% in 2020. During the same time span, India has made significant progress in closing the gender gap, which has decreased from 25.6 to 14.4. However, the issues of educational quality, access, equity, and inclusion remain crucial. Policymakers are also concerned about huge differences in literacy rates between men and women in rural and urban areas.

An autonomous wing of the central Ministry for Human Resource Development, the National Literacy Mission Authority (NLMA) is the nodal agency for the overall planning, management, and financing of the adult literacy endeavor of the government. This agency has taken up various linkages that relate to empowering adults, such as electoral, financial, and legal literacy as part of the convergence scheme with the SDG targets and has also taken proactive steps to align Saakshar Bharat with Sansaad Adarsh Gram Yojana in order to achieve 100% literacy. More than 10 million Saakshar Bharat beneficiaries were also encouraged and mobilized to open bank accounts under the Pradhan Mantri Jan Dhan Yojana (PMJDY), increasing the financial literacy amongst the adult population (Table 12.3).

TABLE 12.3 Trends in Social Sector Expenditure

Items	2009–2010	2013–2014	2014–2015	2015–2016 (RE)	2016–2017 (BE)	2019–2020 (RE)
		As Percentage to GDP				
Total expenditure	28.6	26.7	26.4	29.1	29.5	31
Expenditure on social services of which:	6.9	6.6	6.2	7.3	7.4	7.5
Education	3	3.1	2.8	3.1	3.2	3.1
Health	1.4	1.2	1.2	1.4	1.5	1.28
others	2.5	2.3	2.1	2.8	2.8	3.12

Source: GOI, Economic Survey (2019–2020).

As shown in Table 12.3, "education spending as a proportion of GDP has been hovering around 3% since 2009–2010, compared to 1.4% for health. Uttar Pradesh (67.68), Andhra Pradesh (67.02), Jharkhand (66.41), Rajasthan (66.11), Arunachal Pradesh (65.38), and Bihar are among the states in India with literacy rates lower than the national average (74.04%; 61.80). In this sense, Bihar and Jharkhand have had extremely low reading levels in recent years. On the plus side, Kerala (94.00), Lakshadweep (91.85), and Mizoram (91.33) are the only three states with literacy rates above 90% (ASER, 2021). Despite the Kothari Commission's suggestion that overall government education spending be increased to 6% of GNP by 1986, which was also endorsed by both the original and modified versions of the New Education Policy" (Table 12.4).

TABLE 12.4 Average Annual Drop-Out Rate in School Education (All Categories of Students)

Classes/ Year	Primary			Upper Primary			Secondary			Senior Secondary		
	Boys	Girls	Total	Boys	Girls	Total	Boys	Girls	Total	Boys	Girls	Total
2011–2012	5.89	5.34	5.62	2.13	3.20	2.65	NA	NA	NA	NA	NA	NA
2012–2013	4.68	4.66	4.67	2.3	4.01	3.13	14.5	14.5	14.54	NA	NA	NA
2013–2014	4.53	4.14	4.34	3.09	4.49	3.77	17.93	17.79	17.86	1.48	1.61	1.54
2014–2015	4.36	3.88	4.13	3.49	4.6	4.03	17.2	16.9	17.06	0.25	NA	13.09
2015–2016	4.20	3.19	4.09	3.78	4.8	4.05	17.21	16.88	17.06	NA	NA	NA

Source: MHRD: Educational Statistics at a Glance (2019).

Table 12.4 shows data on dropout rates, "which indicate that while they are decreasing year after year, they remain a significant concern. The yearly dropout rate at the primary level for all categories of pupils has decreased by 1.53% points from 5.62% in 2011–2012 to 4.09% in 2015–2016, thanks to a rise in the number of schools, improved educational infrastructure, and higher enrollment. In contrast, the annual dropout rate in the upper elementary and secondary levels for all categories has climbed to 4.05% from 2.65% during the same time, reaching a peak of 17.06% from 14.54% at the secondary level."

There are various explanations for the high-level dropout rates in India. Some factors can be observed based on research conducted in the state of Rajasthan. These factors include parents' reluctance to allow their children to attend school, as well as a lack of interest in attending school on the part of the children. Other factors include a lack of infrastructure, a shortage of trained teachers, and a high pupil-to-teacher ratio. Crimes against children, as well as the role of child labor, child marriage, and child trafficking, are the primary causes of school dropouts. The administrative responsibilities assigned to primary school teachers also have a negative impact on learning outcomes. Long travel distances, lack of infrastructure facilities, home care of siblings and elderly family members, use of school-age children in cattle farming, lack of access to clean drinking water and sanitary facilities, and the continued use of corporal punishment in schools are the main causes of the decline in inclusivity in the education system. Dropout is a national problem in our country, and there is no one-size-fits-all solution, as not all causes of dropout can be addressed by the school. Rather, all stakeholders – parents, teachers, administrators, the community, and kids – must work together to attain them. Furthermore, when devising intervention measures, students must be considered at the forefront of the problem. Due to the no detention policy, many youngsters dropped out in ninth grade. Many schools had just become a place where children could eat the mid-day meal provided to them, while their education was not paid attention to and degraded. To solve this issue, learning outcomes and training had also been provided to 1.5 million teachers.

12.4 WAY FORWARD TO ACHIEVE QAEI

Education is a significant issue, with India leading the way in the 21st century by implementing innovative flagship programs to achieve the QAEI factor and meet the SDGs on education in a timely manner. As emphasized by the SDG, there needs to be adequate physical infrastructure, safe and inclusive spaces, and nurturing learning outcomes for all, regardless of gender, background, or disability status, in order to achieve the goals. The NLMA has taken various linkages relating to empowerment issues, such as electoral, financial, and legal literacy, as part of the convergence scheme. It has also proactively aligned Saakshar Bharat with Sansaad Adarsh Gram Yojana to achieve 100% literacy. The PMJDY has incentivized more than 10 million Saakshar Bharat beneficiaries to open bank accounts. In recent years, reading levels in Bihar and Jharkhand have been extremely low. On a positive note, the only three states with literacy rates exceeding 90% are Kerala, Lakshadweep, and Mizoram. Despite the Kothari Commission's (1966) suggestion that overall government education spending be increased to 6% of GDP by 1986, which was accepted by both the original and updated versions of the New Education Policy, investments in education have been nowhere near this mark.

In spite of the fact that India has high enrollment and attendance rates, the results from the relevant learning vary across the country. This is a major concern that is frequently emphasized in the context of school education (Economic Survey, 2018). Absenteeism among teachers and a dearth of professionally competent instructors are two underlying causes of low educational quality in the elementary sector. These two factors are still problems that must be addressed (Economic Survey, 2020).

Standardization and regular dissemination of educational data would help in obtaining accurate measurements to monitor the progress in achieving the fourth goal of sustainable development. Additionally, these metrics can be compared across all states and districts to assess the achievements. This will assist in identifying areas where states are falling behind, enabling better focus on policies and initiatives to attain the desired educational outcomes. Therefore, it is imperative to enhance the collection of data from administrative and household sources to achieve this standardization. The government has already assigned nodal ministries and departments for each goal, demonstrating its commitment to long-term development objectives. Education is a cross-cutting issue that spans multiple sectors; it is also a goal in itself, necessitating actions and efforts across various sectors. Furthermore, the allocation of necessary resources for financing education must be rigorously implemented and regularly evaluated.

KEYWORDS

- **child trafficking**
- **corporal punishment**
- **cross-cutting issue**
- **education for sustainable development**
- **gross enrollment rate**
- **interdisciplinary**
- **management education**
- **sustainable development goals**

REFERENCES

Brundtland, G., (1987). *Our Common Future: The World Commission on Environment and Development.* Oxford: Oxford University Press.

Gieryn, T. F., (1983). Boundary-work and the demarcation of science from non-science. *American Sociological Review, 48*(6), 781–795.

NITI Aayog Report, (2021). SDG India Index and Dashboard 2020–21: An Introduction to the Third Edition. https://wgz.short.gy/SDGIndiaIndex.

NITI Aayog, GoI, (2015). *Good Practices Resource Book – Social Sector Service Delivery.* UNDP.

UNESCO Bangkok (2002) Innovations in Non-Formal Education: A Review of Selected Initiatives from the Asia-Pacific Region. Bangkok: UNESCO. Available at: http://www2.unescobkk.org/elib/publications/INFE/p1-14.pdf.

UNESCO International Institute for Educational Planning (IIEP) (2006). Guide for planning education in emergencies and reconstruction: Non-formal education. Paris: UNESCO-IIEP.

United Nations General Assembly, (2015). *Transforming Our World: The 2030 Agenda for Sustainable Development.* United Nations: New York, NY, USA, https://undocs.org/en/A/RES/70/1 (accessed on 20 February 2024).

Wagner, P., (1991). Analyzing social science: On the possibility a sociology of the social sciences. *Discourses on Society, The Shaping of the Social Sciences Disciplines* (pp. 3–22). A Sociology of the Sciences, Kluwer Academic Publishers, Dordrecht, Boston, MA and London.

CHAPTER 13

ICT in Managerial Education for Sustaining Economic Crises: Boons, Challenges, and the Way Forward

SOUMITA MITRA[1] and VIKAS ARYA[2]

[1]*Senior Research Scholar, Department of Humanities and Social Science, IIT (Indian School of Mines), Dhanbad, Jharkhand, India*

[2]*Assistant Professor–Marketing, Rabat Business School, International University of Rabat Salé, Morocco*

ABSTRACT

The strike of the pandemic has transformed the normal flow of life. It has pushed the survival of the common people into the darkness of uncertainty and filled them with a state of mind where they find themselves in total conundrum. The confusion is not only restricted to ensuring the protection of one's life and family members from the claws of this disaster, but the lingering question is how they would do so. Due to the observance of complete lockdown in most countries, companies have started following mass reduction in the workforce. Thus, working individuals were forced to accept a drastic change, either in the form of losing their job or shifting the work setup to the virtual mode, which also suffered from a lot of inconsistencies. Changing the whole workspace into an online mode was not a viable option for most employees, as they did not possess the infrastructural facilities in their homes that would support the shift of their workstations. This work will try to analyze the constraints that entangled the economic condition of the countries and hindered the psychological

Interdisciplinary Approaches in Management Education.
Tanusree Chakraborty, Ramandeep Saini, Rajiv Khosla, Nandita Mishra, and Ashok Natarajan (Eds.)

liberty of the citizens. Some situations in the human world that eventually occur are beyond understanding and control. Therefore, at this juncture, it became evident how possessing technical skills in terms of having proper knowledge of the use of ICT proved to be of great help to many individuals. It has also accentuated why managerial institutes should make students well-equipped with such expertise so that they are able to find maximum employment opportunities for themselves.

13.1 INTRODUCTION

The spread of COVID-19 has created a significant disruption, impacting the lives of individuals in various ways. It has affected every aspect of society, causing extensive damage to physical and mental well-being. Many in the working class have lost their source of income, leading to hardships for their families. The loss of earning capabilities has made it challenging for them to meet basic needs. This has resulted in continuous mental distress and affected their performance at work. The situation has left them feeling helpless and paralyzed. They struggle to provide proper support, including medical aid and daily necessities, to their loved ones during this crisis. Additionally, the education sector has faced significant challenges, requiring a shift towards complete digitalization. However, the situation before the onset of the pandemic was completely different. Education used to primarily rely on face-to-face modes of learning. The sudden shift to digitalization had a profound impact on the entire educational landscape. It brought about a wide range of digitally-driven opportunities, including the adoption of more interactive and collaborative learning strategies. This also led to the modification of teaching materials (Rof et al., 2022). However, embracing change quickly is not always an easy task. Often, any transformation is seen as a negative force. It is because of the various after-effects associated with it that each individual in that concern faces at that point in time. Such thoughts can lead to unfavorable impacts and cause disorganization and misapprehension among the faculties and eminent academicians working in any institute. Therefore, it becomes important to figure out the necessary amendments or changes that must be accommodated for the welfare of everyone present there. In order to incorporate such things, it is essential to understand the organizational needs as well as individual and group perceptions. Thus, for a person who takes the initiative to bring about these transformations, their journey may be inspiring, frustrating, visionary, exhilarating, and fear-inducing. All these would result in developing mixed feelings. However, even after all odds and facing all struggles, the focus must be maintained in keeping a visionary tone and keeping things optimistic. In this context, learning the ways of LMS or Learning Management System would also prove to be helpful in creating a magnification of the change process and enabling an extensive level of observation of the different factors that affect surrounding leadership (Finn III & Feldhaus, 2021). When one views the broader space of learning, it can be understood that knowledge is not just one of the key resources and assets of the organization but is a fundamental tool that enables the effective management of other resources in the organization. Thus, advocating ICT or information and communication technology, it becomes easier for any organization to make steady progress towards future endeavors where increasing technological dependency is necessary to enhance productivity (Funda, 2019). However, during a time of widespread scarcity and anxiety, implementing any change became the most challenging part. In the midst of the pandemic, while many lost their lives battling the disease, there were also those who intentionally harmed themselves due to the serious psychological pressure and depression they faced. Enduring numerous ups and downs in various aspects of life, what exacerbated the situation for all humans was the prolonged period of lockdown for several months. The uncertainty and dizziness about the future made them even more restless. Therefore, such circumstances created a strong need for individuals to be flexible in their approach in order to sustain themselves during this crisis. Those individuals who had prior knowledge of ICT applications found themselves in a much better position than those who were unaware. At this juncture, it is important for educational institutions to equip students with the ability to adapt and be versatile by providing proper training on utilizing ICT in every aspect of their activities.

13.1.1 MAJOR HIGHLIGHTS

- Dependence on the use of ICT;
- Rise of gig economy;
- ICT as a source of communion;
- Importance of acquiring ICT skills;
- Inclusion of ICT in curriculum of managerial education.

13.2 DEVASTATION CAUSED BY COVID-19 AND RISE OF ICT

The arrival and impact of the pandemic in the form of COVID-19 has led to a profound realization in the human mind. It has made us understand that by interfering with and ignoring the damages caused to the environment on a regular basis, we run the risk of being completely overwhelmed. It has also instilled in us a sense of gratitude and the need to acknowledge the bountiful gifts provided by nature.

During the COVID-19 period, the sight of piled boxes carrying deceased bodies and exhausted medical professionals tirelessly serving humanity became all too common. These brave individuals, who themselves became infected with the Coronavirus, ultimately lost their lives while fulfilling their duty. It was an unimaginable event for advanced economies to believe that they would have to face such disastrous situations and suffer heavy losses, both in terms of lives and economic crash. Presently, the conditions have come under considerable control. Still, it can be said that the world had already gone through a really tough time and the sufferings. But the fright and worry have not been totally discarded. This is so because there is yet an increased chance of the new evolved strain of the virus. Thus, the countries have been left with no options, except complying with the same decided rules of keeping social distance and following safety guidelines. The occurrence of this universal calamity has affected not only the medical sector but also created deep craters in the production capacity of the countries. It has also pulled down the quality of the United States, of leading ahead of the other nations and setting an example, due to this challenging situation. All the rest of the nations are now eagerly awaiting to view how this developed nation is going to combat the present disaster. While making constant efforts in coping with this circumstance, there has been a reduction in the influencing and persuasion powers of these countries. Now, the prime focus of most of the nations is to concentrate on keeping a check on the rising cases of COVID-19 within the periphery of the country (Burrows & Engelke, 2020). However, this entire incident could have been managed in a better way, if not fully controlled or avoided. The ill consequences created due to the spread of the corona virus (COVID-19) have left no spheres of the ecosystem un-demolished. Therefore, a little carefulness by all nations could have saved some lives. Caution statements were given many times by the World Health Organization (WHO) and the United Nations (UN). They had previously warned everyone about the possibility of an epidemic that might arise at some point in a year and would cause disastrous effects. Therefore, the continuous neglect of such a serious caveat gave rise to the condition of this calamity in the form of COVID-19. It not only generated health jeopardy but also gave rise to other menaces in the form of crime, virtual bullying, conflicting affairs within the territories, and battles between business groups. In order to stop such instances, it requires added awareness and allocation of funds from the government (Pantucci, 2020). However, if we take a graver look into the outline of the entire episode of COVID-19, then we will eventually find many loopholes that were already present in each and every corner of our social structure. The disaster that has occurred due to the COVID-19 has played the role of a radiography, which has highlighted the drawbacks and flaws that were existing in the social setup of most nations. The situations of deficit that have emerged due to the inadequacy present not only in terms of providing adequate medical facilities to all, but also, it has become evident about the absence of unity in the communal structure. From such gaps, it becomes visible about the showcase of intolerance by the diverse class of individuals. Such persons do not even bother to think about the seriousness of the present situation; rather, they keep themselves busy in trying ways to obtain maximum benefits, in terms of securing important designations or earning status symbols for themselves (Padan, 2020). Thus, it becomes evident that some people are only interested in meeting their personal motives. They, in this process, absolutely ignore the tumult that others are facing in their lives. Here, once again, it gets accentuated that waiting for the opportune moment or the arise of any crisis can push an individual towards the brim of facing extreme challenges. Preparedness can always help a person to remain afloat even in the midst of utmost calamity. Thus, keeping oneself well-versed in ICT skills is a must step for the learners, especially for the ones dealing with managerial discipline. As they are supposed to deal with a large number of people from different backgrounds with various requirements, being efficient in the use of ICT can make their work easy and quick to handle.

13.3 THE TRANSDISCIPLINARY OUTLOOK FORWARDED BY THE MANAGERIAL TEACHINGS TO ADAPT THE SHIFT TOWARDS INDUSTRY 4.0

In order to keep pace with the fast-moving world, we have gradually moved ahead and made arrangements to make our lives more comfortable and easier. Everything has been digitized, a trend that gained wide acceptance from various consumers during the COVID situation due to government-imposed limitations. As a result, when we think of luxurious comforts, options like Uber and Airbnb immediately come to mind. Uber is a digitalized option for hiring rental cars, while Airbnb enables us to arrange and reserve living spaces for ourselves. This system has given rise to a new form of economic setup called the "collaborative," "platform," and "gig" economy. It involves two parties entering into an undertaking through a digitalized way, making it easy to find and book a cab or arrange lodging with the help of a mobile application. It took a long time and involved many transformational steps to reach this level of easiness and flexibility. However, the revolutionary modification in the digital world became popular due to the availability of smartphones among most people and other options like giving ratings to the services they opt for. This not only helps build trust in the services used by consumers but also provides a room for service providers to improve based on the ratings given to them. Thus, the system of promoting shared economic terms between companies has helped create connectedness among all. It enabled the smaller service providers to work jointly under the bigger brand names. These types of shared business teams enabled the sale and exposure of their facilities in a much wider range, thereby reaching maximum people at a time. It also serves to be a viable option for those who prefer not to enter into the old ways of maintaining everything single-handedly (Calo & Rosenblat, 2017). Thereby, such systems enable sharing the burden of risk and not confronting any losses or responsibilities directly. As per OECD or The Organization for Economic Cooperation and Development, an internationally based concern that looks after the work of good and sustainable policy formulation in order to provide better living to all, had recognized three types of ICT skills. The three sections are specialist or advanced, generic, and complementary. The individuals who will be well-versed in the specialist section will be expected to serve in more specific or core areas like working as a web designer. In the second place, employees with expertise in the second will possess a part of the knowledge required to work with ICT. In such cases, individuals will opt for jobs where intricate details of technology will not be required. For instance, the job of an accountant keeper whose work would only be limited to maintaining the account-related affairs through the use of Excel sheets or other allied software serving that purpose. Being well-equipped with the use of technology and knowing its application will always allow a person to not only increase their own skills but also make their absorption in most of the workplace easily done in comparison to others who lack expertise in those areas. Therefore, it becomes very essential on the part of educational institutions to train their students. As per the standard economic model, an indication depicting a lack of adequate candidates for filling up the job vacancies where the skill of ICT would be required is generally marked by salary hikes and an urgent situation for filling up posts by suitable candidates. At this juncture, only those candidates who have already been trained in the use of ICT and are also willing to upgrade their existing talents would benefit. It would also open up the gate of opportunities for those who are willing to learn and adjust as per the requirements of the concern. All this can be made possible through the correct choice of Post-Secondary Education (PSE). On the other hand, colleges and universities should strive to provide learning facilities for students to learn the technicalities of using ICT. Taking effective steps in this regard will bridge the gender gap and maximize employment opportunities for the female population (Lowther et al., 2012). Management schools are no exception to this. Their initiatives in accommodating students with various infrastructural and learning facilities will enable them to develop a dynamic personality and apply for jobs with multidisciplinary profiles. It will also prepare students to face any unforeseen crisis.

13.4 CONSTRAINTS WHICH AFFECTED THE SMOOTH FUNCTIONING OF INDUSTRY 4.0

The disastrous situation created by the spread of the COVID-19 has led to great loss, not only in terms of lives lost but also in deeply affecting the economic conditions of most nations. With the relaxation of previously issued guidelines during the initial days of the predicament, the direct impact was felt on the financial status of nations. The Gross Domestic Product sharply declined by 7.5% in the Netherlands and 7.0% in Germany. This has forced other nations to make unusual and tough choices. The outbreak of the pandemic has had far-reaching effects on the normal functioning of commercial activities, greatly impacting the employees. The complete closure of areas related to daily chores has resulted in a sharp decline in overall customer demand. This has put companies in a "deadlocking situation" where their decisions have become restricted and limited. The atmosphere of panic that was prevalent during this time diminished the trust factor, which had previously remained unaffected by any uncertain circumstances. The reliability of consumers for the products they usually purchased remained intact. As a result, businesses were left in a state of dilemma, unsure whether to clear existing stock or proceed with more production to meet potential increased demand once the situation normalized. However, these were mere assumptions, making it difficult to determine if the current scenario would return to its original state. The extent of damage varied across different sectors of society. For instance, the food industry was not as affected as the supply of season-specific clothing, such as woolen wear. Consequently, such uncertain situations create challenging times for employees. In addition to the difficulty of laying off a large number of employees, organizations also faced the challenge of maintaining operations with a minimal workforce. Maintaining the prescribed norms of physical distance among everyone within the office became a challenging affair. At the same time, there was also the continuous pressure of ensuring safety and a healthy environment within the company. Even if the production-related issues were managed to a certain level, there still lingered a question of how the produced goods would reach or be delivered to the customers (Meester & Ooijens, 2020). In this context, the sector responsible for taking care of the country's sustenance remains an important one, as it is responsible for feeding the entire population. Therefore, the working individuals engaged in this sector are linked with a highly accountable task. However, it is dismaying to find that the people who take on the responsibility of providing food to all are themselves devoid of attaining the minimum requirements in their life. During the lockdown period, migrating workers went through a lot of hardships. Mostly, those who had no proper documents of their identity and who also belonged to minority classes had to endure the maximum suffering during those times. Such occurrences have given birth to a term called "Informal Economy." This encompasses all the commercial ventures where this working population performs their work but are not given any consideration under legal code and in general. The farming zone has the record for the largest number of informal employees. At a global level, it varies from 60%–90% (Elver & Shapiro, 2021). It is a serious issue where the providers of daily meals to the citizens of a country are themselves insecure about their fixed source of income and health, thereby being forced to accept a life full of compromises. Such things call for urgent changes, and care should be taken that those who look after providing such an essential component in the lives of others, in the form of food, are entitled to enjoy a happy and secure life for themselves.

13.5 ROLE OF ICT IN THE ECONOMIC STRUCTURE AND THE NEED FOR ITS IMPLEMENTATION IN MANAGEMENT COURSES

In the modern outlook of life, everything has become a matter of easy access. No one now bothers about the limitations in the form of space, time, and distance. Everything can be bought at any point in time by a person. Besides this, the gradual development in internet facilities and artificial intelligence (AI) has allowed the emergence of ventures like Amazon, Facebook, or Google. All these virtual platforms make all the undertakings of humans possible and quick. The functioning of these Information Technology-oriented facilities depends on a system of applications, which in turn makes use of calculable algorithmic programs. Then, these are transferred to cloud technology, where all the data is saved for either later use or they are used to carry on the whole process of market-related activities. All the virtual measures operate to provide a different level of service to its users. While some work as search engines, the rest help in finding most of the required activities that humans want to have. In a very interesting manner, the virtual modes have acquired their brand names specifically based on the particular facilities they provide. Even the labels that are decided to be displayed for advertising the products that the website offers also bear unique characteristics. From there, the name "Creative Economy" has been derived. In addition to the unique features held by the company's logo, there are other important aspects such as privacy policies and flexible guidelines. These companies act as intermediaries between the sellers and buyers of goods. Although the companies have the right to monitor the movement of ordered goods and exercise complete authority over the entire process, there are also sellers who prefer to showcase a wide range of products through social media platforms like Instagram and YouTube. They aim to provide buyers with a vast selection of options. This type of business expansion, centered around media, is referred to as "platform economy" or "digital platform." It can be said that if the "Age of Industrialization" was focused on the development of manufacturing units, the current transformations in the business markets are primarily based on the digitalization of all resources (Kenney & Zysman, 2016). Presently, the living pattern of most individuals, especially those residing in urban areas, has undergone numerous modifications. They have started adopting a lifestyle based on the use of mobile applications, inspired by practices originally followed by citizens of the United States. This has given rise to a system where individuals can showcase and sell their best-produced items through websites. The range of items that can be sold through such measures is vast, including home-prepared food variants, extra living space, daily family articles, personal expertise, and even transportation facilities.

This technique was subsequently adopted by brands like Uber and Task-Rabbit. While in some cases, this shared business outlook was created to earn profits, in other cases, it served charitable purposes and allowed for the economic use of leftover items. Examples of service providers not directly oriented towards gains include Zipcar and Freecycle. Thus, this mechanized form of commerce rapidly grew in order to survive in situations of market downturn. During this time, individuals also realized the importance of reusing old items in different ways. To transform this entire operation into a successful venture, the online network has been of great help. Moreover, with the aid of mobiles equipped with innovative features and the availability of cloud computing, it has become easier for individuals who wish to accomplish their tasks with assistance from other sources (Sprague, 2015). This allows buyers to complete their work more efficiently, while the service provider earns a share of the profit from the services provided. Therefore, with the continuous advancement in technology, a transaction system has emerged where a connection is established between an order placed and the goods or services supplied to the buyer (Waas et al., 2018). Thus, it can be observed that digitalized platforms are gradually becoming the new norm of the present system. Although the pandemic has made it a necessity and the only viable option, its wide acceptance has turned it into a popular medium. In this regard, knowledge of ICT is an important requirement for new generation job seekers.

13.6 LIMITATIONS OF THE DIGITAL WORLD OR ICT IN THE ERA OF INDUSTRY 4.0

Technology-based or digitized ways of accessing services and goods are gradually becoming a part and parcel of everyday life for common people. The increased dependency on such service-providing companies has opened up numerous employment opportunities for young individuals. In developed nations like America, where the population is constantly seeking work options that can accommodate their complex lifestyles, there has been a shift away from the traditional employer-employee relationship. Technical innovation, coupled with changes in economic conditions, has led to the rise of self-reliant contractors and contractual working staff. In the United States, over 16% of the population prefers flexible terms in their workplace, maintaining flexible connections between the main employer and subsidiary employees (Hicks, 2017). When discussing the widespread use of technological progress, the term "InsurTech" comes to mind, referring to the computerized systems used by insurance companies. Additionally, "FinTech" or "Financial Technology" is a faster way of delivering services related to financial activities. As this system has progressed, chances have been reduced for people to bring innovations in the banking and security sectors. All the technical expansion has already been made possible through the implementation of AI or "artificial intelligence." In addition, when insurance providers use such a system, they utilize "talkbot" or "chatbot" in the form of neutral networks. Talkbots or chatbots provide quick replies to user questions, while neutral networks help identify fraudulent activities. Blockchain is also a method through which all monetary transactions can be stored and saved with the aid of computer systems for future use. It is also helpful in reducing the chances of forgery and increasing reliability, ultimately improving performance (Neale et al., 2020). Thus, a large section of the workforce is slowly being replaced by machines. Positions and tasks that were previously handled by humans are now controlled by machinery. This has reduced the chances of errors that may occur when humans perform those jobs. With the use of digital assistance, results or outcomes from any work are derived quickly and flawlessly. But, as the line adapted from the play "Macbeth" by William Shakespeare says, "Security is mortal's chiefest enemy." This line becomes very relatable to the present situation of the world. While the advent of technology and massive success in the field of technology has made our lives easier, we also become more vulnerable to cases of falsification, where a person might lose their entire bank account balance. On one side, the digital world has enabled easier connectivity with anyone, at any time. On the other side, it has made all our personal details easily available. One moment of mistake while sharing one's own bank details might open up the way for scammers to access all information within seconds. Things turned out to be very strenuous, especially during the outbreak of COVID-19 when mobility became restricted. For most individuals, online purchase of daily necessities was found to be the best option. They preferred goods to be supplied to their doorsteps instead of going outside. They also chose this alternative to reduce the risk of getting infected with the disease. However, cases were reported where people either lost their money or received defective items. Thus, cases like this may become the reason for added stress instead of reducing workload and ensuring safety, while increasing one's dependence on a digitalized way of life.

13.7 A RESPONSIVE APPROACH OF INFORMATION AND COMMUNICATION TECHNOLOGY (ICT) TO COUNTER THE ONSLAUGHTS OF COVID-19

The pandemic has forced everyone to undergo a re-examination of their living habits. It has prompted us to reassess our previous ways of interacting with other beings on this planet (Junker, 2020). The main sectors of society that consume the highest amount of power include industrial belts, transportation, farming, corporate houses, institutions, and daily household chores. Apart from agricultural and domestic work, most other areas that were closed due to government protocols to prevent the further spread of the disease experienced a significant change and reduction in overall power consumption. For instance, the need to import petroleum decreased by 37%, from PKR 170 billion in February to PKR billion in March. This led to adjustments in government budget plans and improved the country's foreign exchange condition. Furthermore, the continuous closure of many activities in nations resulted in rapid and positive changes in the ecosystem. It brought about a decline in the amount of harmful greenhouse gas emissions in the atmosphere, resulting in a refinement of the global temperature level. According to news obtained from various sources, it was highlighted that a noticeable recovery was observed at the atmospheric level due to the slowdown of human activities. Advancements were observed in both the climate and the aquatic ecosystem, while also reducing expenses incurred in projects aimed at mitigating and adapting to changes on the planet (Hassan, 2020). To illustrate how such practical changes were enabled, the case of Malaysia can be taken into consideration. Malaysia is one such country where people belong to mostly upper-middle-class income-earning groups. However, the way they combated the crisis was praiseworthy. They had been engaged in being prepared for facing any situation of crisis long before the actual incident occurred. In working towards this motive, they collaborated with the international health regulations (IHR), 2005, to strengthen their capability to avoid such occurrences in the future. Even if they occur, they would be able to identify the spread of disease at an early stage and take timely actions to save maximum lives. Before the calamity actually hit the world, Malaysia was one nation that had already prepared themselves with the aid of governmental concerns and health officials to tackle such an obstacle (World Health Organization, 2020). Apart from the above-mentioned examples, one of the most optimistic transformations that took place during this time was in the sphere of Information Communication Technology. It acted as a center of communion for people to emerge from isolation and find common ground to support the well-being of others. It proved to be a platform where everyone belonging to diverse backgrounds and opinions came together, enabling them to share their creativity with people from different parts of the world (Yang et al., 2020).

13.8 CONCLUSION

Change or transformations are the most obvious and unavoidable part of human civilization. We need to make amendments or necessary adaptations as per the changes occurring, lest we should remain left out. Thus, the digital form of lifestyle, which was slowly advancing in the world market, took a massive form, thereby becoming a part and parcel of everyday life for all individuals. At this point, it becomes important for Business Schools to train their students with ICT skills. Along with that, stories of inspirational leadership from other nations should motivate them to follow suit. ICT, in this difficult situation, has not only helped connect people but has also advocated every possible measure to provide everyone with necessities and emergency support, thereby broadening the chain of connectivity amongst all.

13.8.1 POINTS TO PONDER ON

1. **Predominant Skill:** Acquiring expertise in ICT has become one of the prerequisites to enhance one's chances of employability.
2. **Inclusion of ICT in Managerial Curriculum:** The responsibility of B-schools lay in training students with knowledge of digital learning, thereby making it an integral part of the curriculum.
3. **Source of Connectivity:** Due to the impact of the pandemic, it was the only means of staying connected with one another.
4. **ICT as a Next Gen Factor:** To keep pace with the changing needs of the job market, it became important to learn ICT skills.

KEYWORDS

- **artificial intelligence**
- **COVID-19**
- **employability**
- **gig economy**
- **industry 4.0**
- **information and communication technology**
- **managerial teachings**

REFERENCES

Burrows, M., & Engelke, P., (2020). *What World POST-COVID-19?: Three Scenarios.* Atlantic Council. Retrieved from: www.jstor.com/stable/resrep24634 (accessed on 20 February 2024).

Calo, R., & Rosenblat, A., (2017). The taking economy: UBER, information, and power. *Columbia Law Review, 117*(6), 1623–1690. Retrieved from: www.jstor.org/stable/44392959 (accessed on 20 February 2024).

Elver, H., & Shapiro, M., (2021). Violating food system workers' rights in the time of COVID-19: The quest for state accountability. *State Crime Journal, 10*(1), 80–103. Retrieved from: https://heinonline.org/HOL/LandingPage?handle=hein.journals/stecrjl10&div=7&id=&page= (accessed on 20 February 2024).

Finn, III. E. W., & Feldhaus, C., (2021). Perception in leading change. *The Journal of Technology Studies*, *47*(1). 24–37. Retrieved from https://www.jstor.org/stable/10.2307/48657933 (accessed on 20 February 2024).

Funda, V. N., (2019). *Impact of Information Technology on Knowledge Management at a Selected University of Technology* (pp. 1–154). Cape Peninsula University of Technology. Retrieved from: https://core.ac.uk/download/pdf/322968397.pdf (accessed on 20 February 2024).

Hasan, M. H., (2020). Brighter side of the global threat. *Policy Perspectives*, *17*(1), 176–178. Retrieved from: https://www.scienceopen.com/document_file/04f0ec71-8208-436b-8454-f047c3d93c74/ScienceOpen/polipers.17.1.0175.pdf (accessed on 20 February 2024).

Hicks, M. S., (2017). Flexible jobs give workers choices. *Monthly Labor Review*, 1, 2. Retrieved from: www.jstor.org/stable/10.2307/90007796 (accessed on 20 February 2024).

Junker, Y. A., (2020). On COVID-19, U.S. uprisings, and black lives: A mandate to regenerate all our relations. *Journal of Feminist Studies in Religion*, *36*(2), 117–129. Retrieved from www.jstor.org/stable/10.2979/jfemistudreli.36.2.09 (accessed on 20 February 2024).

Kenney, M., & Zysman, J., (2016). The rise of the platform economy. *Issues in Science and Technology*, *32*(3), 61–69. Retrieved from https://www.jstor.org/stable/24727063 (accessed on 20 February 2024).

Lowther, D. L., Inan, F. A., Ross, S. M., & Strahl, J. D., (2012). Do one-to-one initiatives bridge the way to 21st century knowledge and? *J. Educational Computing Research*, *46*(1), 1–30. doi: http://dx.doi.org/10.2190/EC.46.1.a.

Meester, J., & Ooijens, M., (2020). *COVID-19 Impact on the Value Chain — Conceptual Paper.* Clingendael Institute. Retrieved from www.jstor.org/stable/resrep25670 (accessed on 20 February 2024).

Neale, F. R., Drake, P. P., & Konstantopoulos, T., (2020). InsurTech and the disruption of the insurance industry. *Journal of Insurance Issues*, *43*(2), 64–96. Retrieved from www.jstor.org/stable/10.2307/26931211 (accessed on 20 February 2024).

Padan, C., (2020). *The Coronavirus Crisis and the Opportunity for Social Transformation.* Institute for National Security Studies. Retrieved from www.jstor.org/stable/resrep27801 (accessed on 20 February 2024).

Pantucci, R., (2020). Key questions for counter-terrorism post-COVID-19. *Counter Terrorist Trends and Analyses*, *12*(3), 1–6. Retrieved from www.jstor.org/stable/10.2307/26915443.

Rof, A., Bikfalvi, A., & Marques, P., (2022). Pandemic-accelerated digital transformation of a born digital higher education institution. *Educational Technology & Society*, *25*(1). 124–141. Retrieved from https://www.jstor.org/stable/10.2307/48647035 (accessed on 20 February 2024).

Sprague, R., (2015). Worker (Mis) classification in the sharing economy: Trying to fit square pegs into round holes. *ABA Journal of Labor & Employment Law*, *31*(1), 53–76. Retrieved from https://www.jstor.org/stable/10.2307/26410781 (accessed on 20 February 2024).

Waas, B., Pavlou, V., & Gramano, E., (2018). Digital economy and the law: Introduction to this special issue. *Work Organization, Labor & Globalization, 12*(2), 7–11. Retrieved from www.jstor.org/stable/10.13169/workorgalaboglob (accessed on 20 February 2024).

World Health Organization, (2020). *MALAYSIA Strong Preparedness and Leadership for a Successful COVID-19 Response*. Retrieved from: www.jstor.org/stable/resrep28089 (accessed on 20 February 2024).

Yang, S., Fichman, P., Xiaohua, Z. X., Sanfilippo, M., Li, S., & Fleischmann, K. R., (2020). The use of ICT during COVID-19. In: *83rd Annual Meeting of the Association for Information Science & Technology*. doi: 10.1002/pra2.297.

CHAPTER 14

An Analysis of Online Education During a Pandemic Situation, with Reference to India

VISHAL SRIVASTAVA, AMRINDER SINGH, and PADMAKAR SHAHARE

JAIN (Deemed to be University), Bangalore, Karnataka, India

ABSTRACT

This research delves into the intricacies of the current online education (COE) system, exploring its potential as a viable alternative to traditional face-to-face education, particularly in the context of pandemics like the COVID-19 crisis, focusing on the challenges and opportunities it presents for large populations in developing nations such as India. Through the collection and analysis of opinions and perceptions from 85 students and their parents, employing non-parametric methods like the chi-square test, the study uncovers a nuanced understanding of online education's role in epidemic situations. While there is a general consensus on the advantages of online education during crises, the study reveals significant complexities within the current online education formats, primarily stemming from the lack of desired user interaction. The findings highlight various limitations, including digital technology shortages and less interactive platforms, which ultimately impede online education from being a full-fledged replacement for traditional offline education. Consequently, the study underscores the need for policy reforms to address these challenges and enhance the effectiveness of online education, particularly in densely populated developing countries like India, where such reforms could have a substantial impact on educational outcomes and access.

Interdisciplinary Approaches in Management Education.
Tanusree Chakraborty, Ramandeep Saini, Rajiv Khosla, Nandita Mishra, and Ashok Natarajan (Eds.)

14.1 INTRODUCTION

People will always remember the year 2020 due to the global outbreak of the communicable coronavirus (COVID-19). Within three months, the virus infected 1,205,801 people worldwide, with 64,973 reported deaths as of April 4th, 2020 (source: https://www.worldometers.info/coronavirus/). This pandemic has caused unprecedented pain and affected every individual on Earth. Many people were confined to their homes, and numerous social and professional activities were disrupted by COVID-19 (Bosan et al., 2020). Education is one of the sectors greatly impacted by this crisis. Education plays a crucial role in personal growth and stability (Brandt et al., 2020). Face-to-face or offline education has been widely recognized as the most comprehensive and effective method. It has been proven to have a positive impact on learners' knowledge, attitudes, and behavior (Esfandari et al., 2020). In the face of the COVID-19 epidemic, social distancing has been considered the most effective measure to prevent the spread of the virus (Levard & Lo, 2020). In the traditional offline education model, it is very difficult to maintain the necessary social distancing. Therefore, most scholars and policymakers recommend using online education as a replacement for offline education for learners (Taser, 2020). To prevent virus spread in mass, various countries have asked their universities to focus more on online courses for their students (https://www.theguardian.com). This research assesses the readiness to use online education as a replacement for offline education in a developing economy like India. However, the year 2001 is considered a milestone for online courses when the Massachusetts Institute of Technology, Oxford, Yale, and Stanford University launched large-scale online programs and massive open online courses (MOOCs) (Jokivirta, 2006). Online education is now preferred by most students to develop their professional and technical skills. A research agency, researchandmarkets.com, has projected the online education market size to be US $187.877 billion in 2019, and it is expected to reach US $319.167 billion in 2025 with a CAGR of 9.23% (Global Online Education Market – Forecast 2020 to 2025). Meanwhile, according to technavio.com (online-education-market-in-India-share-2018), India's online education market is projected to reach a CAGR of 20% to $20 billion by 2022.

The purpose of this research chapter is to investigate the acceptance and limitations of online education in India, particularly during a communicable disease crisis. Additionally, the research aims to assess the impact of identified challenges on the applicability of online education.

14.1.1 RESEARCH OBJECTIVES

- To determine whether individuals exhibit indifference towards the suitability of online education during epidemic situations.
- To assess the perceptions of different users regarding the suitability of online education.
- To identify various challenges in effectively implementing online education based on predetermined parameters and sub-criteria.
- To identify and validate the significance of various research-based dimensions of the online education system on user experience and perceptions.

Succeeding sections of this chapter elaborate on the literature and variables related to success, acceptance, and challenges in online education. Additionally, this section also discusses the parameters established to measure the suitability of any social health management activity during an epidemic situation.

14.2 LITERATURE SURVEY

To comprehend the applicability of online education during a communicable disease outbreak, this study examines the challenges and significance of success factors in online education. Numerous research studies have been conducted on this subject. The literature review of previous research on online education encompasses institutional management, learning environment, technical infrastructure, instructional design, and support services (Chejwindakaran, Suvanantachote, & Therungchaudry, 2013), technology demonstration (Volery & Lord, 2000), as well as instructor and student motivation (Selim, 2007), which have been identified as critical success factors (CSF). However, like any other model, the current model of online education is not entirely compatible with all learners. There are still numerous challenges in online education with its current design and implementation (Martínez-Martínez et al., 2019). Based on previous research conducted in developing countries, major challenges in online education include technological competency and proficiency of students and instructors (Gillett-Swan, 2017; Mashile & Pretorius, 2003), psychological barriers (Ridenour, Simmons, Ilg, & Place, 2005), as well as engagement issues, lack of excitement, anxiety, and fear (Conrad, 2002). Succeeding sections discuss the research methodology and strategy adopted to achieve the research goal.

14.3 METHODOLOGY

Considering previous research works, which are outlined in the previous section, the research adopted an exploratory and descriptive research methodology. The study collected data about the challenges associated with online education and its applicability in the event of an epidemic in developing countries. The study collected the opinions of direct beneficiaries of online education, i.e., school and college students. Along with this study, opinions of parents of primary class, junior class, and secondary class students were also sought, as these parents are very closely connected with their children's education system. To measure the applicability of online education during an epidemic situation (AOP), five indicators – accessibility (AOP1), adaptability (AOP2), convenience (AOP3), efficacy (AOP4), and security (AOP5) – have been used. While the acceptance of online education (AOE) has been measured on technology scalability (AOE1), instructor quality (AOE2), student motivation (AOE3), student support services (AOE4), learning platforms (AOE5), program framework and cost (AOE6), and students' technical knowledge (AOE7). Finally, experienced challenges (COE) of current online education (COE) practices have been measured on access to digital technology (COE1), use of digital content (COE2), convenience in virtual classrooms (COE3), and instructor competence (COE4) parameters. Each variable of experienced/perceived challenges was further measured on specific indicators, as given below:

1. **Access of Digital Technology (COE1):** Lack of awareness (COE11), anxiety or fear of technology (COE12), lack of technical exposure (COE13), improper communication practices (COE14).
2. **Handling Digital Content (COE2):** Technology handling (COE21), student motivation level (COE22), lack of comfortability

(COE23), lack of in-person experience (COE24), customization (COE25), feel of loneliness (COE26).

3. **Comfortability with Virtual Classes (COE3):** Difficulty to come out from comfort zone (COE31), lack of connectivity with peer group (COE32), inability to establish virtual connection (COE33).
4. **Instructor Capability (COE4):** Lacking equipment (COE41), lacking technical efficiency (COE42), lacking online teaching skills (COE43), inability to connect with virtual student (COE44).

The internal consistency exists in all the mentioned scales. The reliability test was conducted using SPSS-20. The reliability test statistics are given in Table 14.1.

TABLE 14.1 Reliability Test Statistics

Scale	Cronbach's Alpha	Number of Item
Applicability of online education (AOP)	0.68	5
Acceptance of online education (AOE)	0.77	7
Challenges in current online education (COE)	0.63	4
Access of digital technology (COE1)	0.66	4
Handling digital content (COE2)	0.71	6
Comfortability with virtual classes (COE3)	0.58	3
Instructor capability (COE4)	0.66	4

The research used the Google Forms platform to connect with respondents. A structured questionnaire tool was used to collect the opinions of the respondents. The questionnaire link was distributed through email, SMS, Facebook, and the WhatsApp Messenger app. A total of 105 responses were received from the population, out of which 85 questionnaires qualified for further analysis.

14.3.1 RESEARCH FRAMEWORK

The research used the population proportion estimation method (Terpstra, 2004; Narjis & Shabbir, 2018; Zamanzade & Wang, 2017) for data analysis to evaluate the proposed model (Hair, 2014). According to the literature survey, the study proposed the model shown in Figure 14.1.

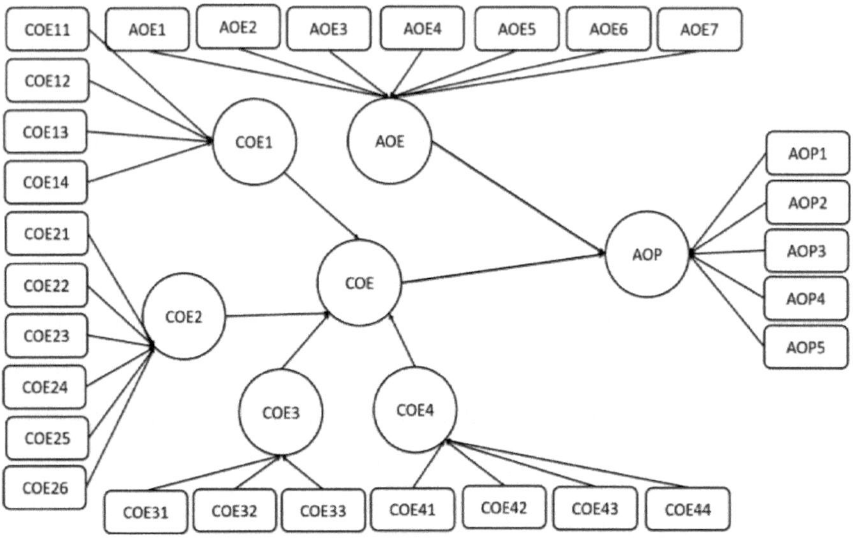

FIGURE 14.1 Challenges in online education in developing country.

The study measured users' opinions regarding the acceptance, challenges, and usability of online education during an epidemic. Thus, the hypotheses formulated to be tested are:

- **H_{01}:** The applicability of online education (AOPs) is not linked to its critical success factors (AOEs).
- **H_{02}:** The challenges of online education (COEs) are not related to the applicability of online education (AOPs).
- **H_{03}:** The challenges of accessing digital technology are not linked to being unaware (COE11), feeling anxious or fearful of technology (COE12), lacking sufficient technical experience (COE13), or failing to communicate effectively (COE14).
- **H_{04}:** Technology handling (COE21), low motivation (COE22), lack of comfortability (COE23), lack of in-person experience (COE24), lack of individual approach (COE25), and feeling of loneliness (COE26) are not associated with the challenge of accessing online education content.
- **H_{05}:** Resistance to change (COE31), lack of connectivity with peer groups (COE32), inability to establish virtual connections

(COE33) are not responsible for users' inconvenient engagement with online education.

- **H_{06}:** The acceptance of online education (AOE) does not pose any challenges for instructors due to the non-availability of equipment (COE41), lack of technical proficiency (COE42), lack of online learning skills (COE43), and inability to engage with virtual students (COE44).

14.3.2 SAMPLE

The study collected data from various segments of direct and indirect users of education services. The entire population is classified into four groups, namely:

- Parents of primary, junior, and secondary class students;
- School going students;
- College going students; and
- Advanced education aspirants.

To obtain a comprehensive perspective, data is collected from various locations in India. Google Forms were utilized to ensure that only internet-enabled users participated in the survey, reducing the likelihood of sampling errors. It is assumed that respondents are accessing online education platforms through their smartphones or desktops, which are equipped with an internet connection, as they would be unable to complete Google Forms otherwise. The distribution of the aforementioned segments in the sample is provided in Table 14.2.

TABLE 14.2 Sample Composition

SL.	Segment	Segment Code	Count	Percentage (%)
1.	College going students	S1	46	54%
2.	School going students	S2	11	13%
3.	Parents	S3	17	20%
4.	Advanced studies or others	S4	11	13%
	Total		85	100%

14.3.3 INSTRUMENT

The structured questionnaire used consists of four sections. The first section collects information about the respondent's association with online education platforms.

The second section of the questionnaire measures the challenges perceived or experienced by respondents (COE) in using an online platform for education. This section measures challenges across four variables as prescribed by previous research (Gillett-Swan, 2017; Mashile & Pretorius, 2003; Ridenour, Simmons, Ilg, & Place, 2005; Conrad, 2002). Each dimension is further measured by several indicators, as mentioned in the flow chart given in Figure 14.2.

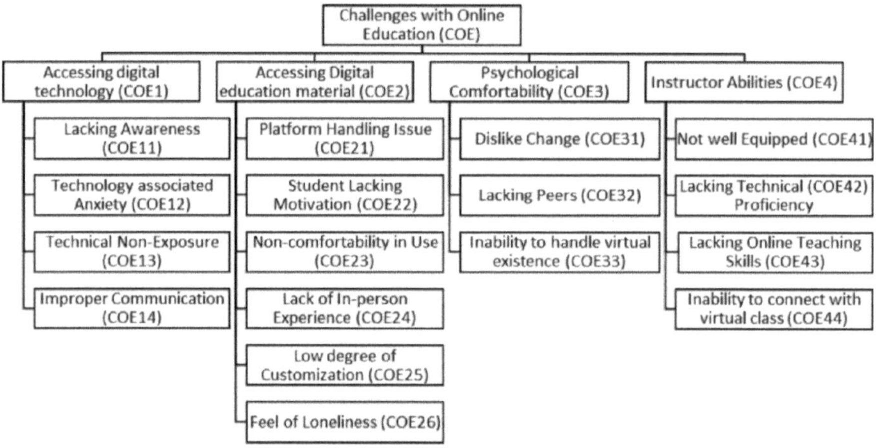

FIGURE 14.2 Perceived/experience challenges in online education.

The respondents have been asked to provide their opinion regarding their experience with the challenges presented. The research utilized a three-point scale consisting of 'yes, no, and can't say.'

The third section of the questionnaire is designed to record the perception or opinion of the respondents regarding the importance of parameters prescribed by previous research works (Cheawjindakarn, Suwannatthachote, & Theeraroungchaisri, 2013; Volery & Lord, 2000; Selim, 2007) in establishing the AOE among users. These parameters include technology applicability of online education (AOE1), online instructor quality (AOE2), maintain user's motivation (AOE3), student support services (AOE4), teaching platform (AOE5), program framework and cost (AOE6), and technical knowledge of the user (AOE7). The importance of these parameters is assessed using a three-point categorical scale – 'yes, no, and can't say.'

The fourth and final section of the questionnaire is designed to measure the opinion towards the use of online education during pandemic situations, such as COVID-19, where social isolation is highly recommended (Long, 2020). This section collects the respondents' opinion in favor of online education based on prescribed parameters that are considered essential for any activity aimed at ensuring social health (Kelley & Hurst, 2006). These parameters include accessibility (AOP1), customization (AOP2), convenience (AOP3), effectiveness (AOP4), and safety (AOP5). Once again, a three-category scale (Favorable, Unfavorable, and Can't say) is used to measure the responses.

14.4 FINDINGS AND DATA ANALYSIS

First of all, the study used the population proportion estimation method to determine the association of various user segments with the identified parameters. The analysis strategy is provided in Figure 14.3.

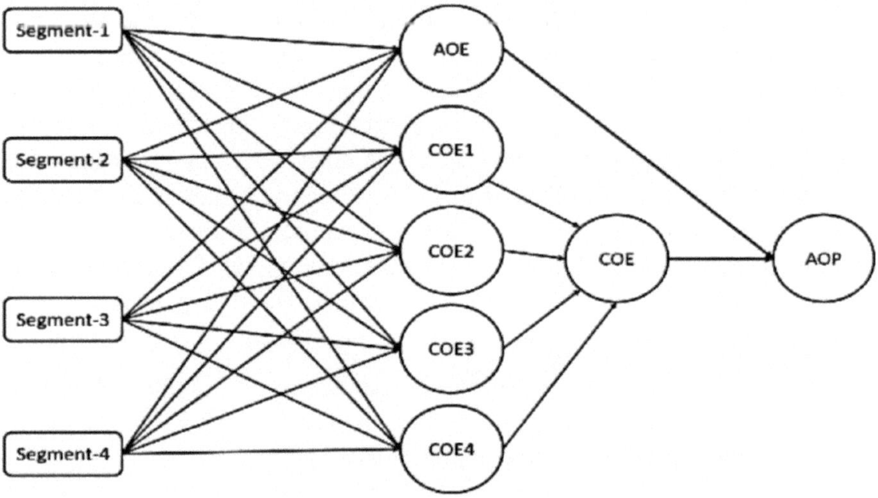

FIGURE 14.3 Analysis strategy for population proportion estimation method.

14.4.1 APPLICABILITY OF ONLINE EDUCATION

First of all, the applicability of online education during a pandemic situation was assessed. As mentioned earlier, the applicability was evaluated based on five parameters across three categories: Favorable, Unfavorable, and Can't say. To determine whether the sample had any indifference towards the applicability of online education during the pandemic situation, a chi-square test was conducted. The P-value of the Chi-Square test was found to be 0.00032. This p-value indicates that the population is not indifferent towards online education during a pandemic situation.

The study also examined the experience and perception of different user segments regarding the applicability of online education in terms of Accessibility, Customization, Convenience, Effectiveness, and Safety.

14.4.2 CHALLENGES WITH ONLINE EDUCATION

Almost all user groups have found that online education is more accessible (93%), customized (91%), and convenient (88%) for educating during epidemic conditions. However, some professionals and individuals engaged in advanced studies or other fields hold different opinions. Approximately 18% of professionals or individuals from other categories reacted adversely to all three criteria, expressing doubts about the effectiveness of online education. Particularly, the majority of school-going students, parents, and individuals engaged in advanced studies or other fields do not find it effective. Additionally, it has been observed that people do not feel entirely safe about online education (19%). One of the main reasons for this concern is the potential adverse effects on the eyes and brain caused by mobile devices and laptops. When people were asked about the challenges they faced in online education, they highlighted the lack of technical access (61%), lack of technical knowledge and skills (58%), and difficulties in adapting to technology (69%). However, to a lesser extent (46%), untrained teachers were also held responsible. Further analysis of the data segmented by user groups revealed the following:

- Advanced studies aspirants lack access to digital technology (73%), although they have the least issue with knowledge/skill in using digital education material (45%).
- Almost all users, except advanced studies aspirants, face knowledge/skill issues in using digital education material (i.e., 59% of college students, 55% of school students, and 65% of parents).

- Surprisingly, it is found that almost all segment users lack comfortability with digital platforms.
- Majority of parents (59%) and advanced studies aspirants (55%) are not happy with the quality of instructors.

  To gain a deeper understanding of these challenges, the study measured all four parameters using sub-parameters. The observations on these sub-parameters are provided below. In order to analyze the challenges related to accessing digital technology, the study sought the opinions of respondents on four parameters: lack of awareness, anxiety or fear of technology, technical risk, and lack of appropriate communication practices.
- It was discovered that nearly all segments face challenges in accessing technology due to insufficient awareness. However, college students exhibit relatively higher levels of awareness (65%).
- Amazingly, various segments are facing challenges in accessing digital technology, but they do not experience significant anxiety or fear when using it. However, this issue is somewhat reported among college students (54%) and parents (65%).
- It is observed that the second factor contributing to the challenge of accessing digital technology is the lack of technical exposure, which affects almost all user segments (70% of college students, 82% of school students, 73% of advanced studies or others, and 82% of parents).
- Moreover, the survey data also suggests that the majority of respondents (74% of college students, 91% of school students, 64% of advanced studies or others, and 94% of parents) face challenges in accessing digital technology due to improper communication practices.
- To understand the problems related to knowledge or skills in using digital education content, respondents were asked questions about the ease of use of technology, their motivation level, their solidarity with technology, human touch, personal level, and fear of helplessness. The observations are given below.
- The outcomes suggest that there are several challenges in accessing digital technology among the population. These challenges include the inability to handle technology (81%), lack of in-person experience (83%), lack of comfort with digital content (75%), and low level of customization (75%).

- Additionally, low user motivation and feelings of loneliness or helplessness are also important psychological challenges in accessing digital technology.

An analysis of the observations based on user groups reveals that almost all user groups face challenges in using digital learning materials. Furthermore, the enthusiasm of students' parents towards digital education is relatively low compared to other users (82%). Except for college students (70%), all other users struggle with studying on digital platforms. The lack of human interaction in the current digital education system is found to be challenging for all user groups. While users in primary to college-level classes generally follow a specified curriculum and have less demand for customization, professionals, and those aspiring for advanced education (82%) are not satisfied with the less flexible format of COE. However, it is worth noting that not all user segments feel isolated with the digital education format, which contributes to its acceptance.

The 69% of respondents complained about the lack of connectivity or comfortability with online education. To understand the problem, this research measured the issue on three parameters: inertia for change (C31), lacking connectivity with peers (C32), and inability to connect with the virtual world (C33). It was found that all three factors are affecting the respondents. Inertia for change was found to be high in college students (83%) and advanced studies aspirants/other categories (91%) (Table 14.10). However, school-going students were found to be most comfortable with change (55%). However, in the online education platform, all learners feel disconnected due to being loosely connected to peers. This may also be due to the absence of interpersonal interaction. It has also been observed that the medium of COE is not capable enough to connect learners with their virtual existence. All user groups, except school-going students (64%), complained about the inability to connect to the proposed virtual world during learning. Instructors are considered the most important element of an education system. The study did not find a clear consensus on the quality of instructors. Among the respondents, 46% expressed dissatisfaction with the competence of instructors, while 27% had no complaints and 27% were unable to provide an opinion. Instructors are considered the most crucial component of the education system. The study revealed that there was no definitive agreement on the quality of instructors. Around 46% of the respondents reported a lack of competence among instructors, while 27% had no concerns regarding their qualifications. Around 27% of the respondents were unable to provide an opinion on the matter. However, when asked about their views on the four key areas of instructor quality, the respondents expressed agreement on the instructors' deficiencies in all four aspects. When user groups were questioned about the availability of necessary equipment with instructors, it was found that 55% of school-going students were not particularly concerned about equipment availability, whereas 82% of advanced education aspirants expressed greater concern. However, almost all users reported a lack of technical proficiency in instructors, as well as a lack of online learning (OLL) skills and an inability to connect with virtual students.

14.4.3 CRITICAL SUCCESS FACTORS (CSFS)

Furthermore, the study verified the importance of the dimensions considered crucial for AOE, as prescribed by previous research. These dimensions include Technology Applicability of online education (AOE1), online instructor quality (AOE2), maintenance of user's motivation (AOE3), student support services (AOE4), teaching platform (AOE5), program framework and cost (AOE6), and technical knowledge of user (AOE7) (Cheawjindakarn, Suwannatthachote, & Theeraroungchaisri, 2013; Volery & Lord, 2000; Selim, 2007). As mentioned above, these parameters are measured on a five-point scale. It was found that the majority of the population (90%) considers almost all the parameters to be important. However, a few insignificant inverse responses were also found. To gain further insight into the population's perception and opinion towards these parameters, the researchers analyzed them segment-wise. Surprisingly, it was found that all segments considered all the mentioned criteria equally important for the AOE. Additionally, all users (100%) agree that the quality of instructors, motivation level of students, student services, and learning platforms are also very important for the AOE. Moreover, the technical know-how of the students is also deemed crucial for the AOE.

14.4.4 HYPOTHESIS TESTING

- **H01: The applicability of online education (AOPs) is not associated with its CSFs (AOEs).**

  The test values and p-values of the 35 chi-square tests between 7-AOEs and 5-AOPs are given in Tables 14.3 and 14.4, respectively.

TABLE 14.3 MAOEs vs. AOPs Chi-Square Test Out-Come

CSFs	Applicability of Online Education				
	AOP1	AOP2	AOP3	AOP4	AOP5
AOE1	08.412	4.692	20.073	05.262	06.832
AOE2	12.470	10.416	14.031	09.606	11.843
AOE3	11.319	10.455	26.652	17.663	11.592
AOE4	11.080	07.426	34.979	08.891	07.357
AOE5	09.071	07.426	23.106	07.457	03.195
AOE6	14.988	22.901	29.861	17.422	11.865
AOE7	10.632	07.439	29.776	06.351	07.645

TABLE 14.4 AOEs vs. AOPs Chi-Square Test p-Values

AOEs	Applicability of Online Education				
	AOP1	AOP2	AOP3	AOP4	AOP5
AOE1	0.493	0.86	0.66	0.949	0.869
AOE2	0.052	0.108	0.081	0.294	0.158
AOE3	0.254	0.315	0.01*	0.126	0.479
AOE4	0.27	0.593	0.00*	0.712	0.833
AOE5	0.431	0.589	0.03*	0.826	0.994
AOE6	0.091	0.006*	0.003*	0.134	0.457
AOE7	0.302	0.592	0.003*	0.897	0.812

*Statistically significant.

The test statistics rejected the null hypothesis of no association between SFs and their CSFs, except for AOE6. In these cases, the association of convenience (AOP3) was not found with the motivation level of students (AOE3), student support services (AOE4), teaching platform (AOE5), program framework operations and cost (AOE6), and technical knowledge of students (AOE7). The p-values for these tests were found to be lower than the critical p-value (0.05), respectively – 0.009, 0.000, 0.027, 0.003, and 0.003. Additionally, customization (AOP2) was also not found to be associated with program framework operations and cost (AOE6), p-value = 0.006.

- **H_{02}: The challenges associated with online education (COEs) are distinct from the applicability of online education (AOPs).**

  The study conducts 20 chi-square tests between four COEs and five AOPs. The chi-square values calculated are presented in Table 14.5, and the corresponding p-values are provided in Table 14.6. The test data demonstrates the relationship between various identified challenges and the applicability of online education. However, the challenge posed by an untrained instructor (COE4) was not found to be associated with the customization (AOP2) and effectiveness (AOP4) of online education.

TABLE 14.5 COEs vs. AOPs Chi-Square Test Out-Come

COE	Applicability of Online Education				
	AOP1	AOP2	AOP3	AOP4	AOP5
COE1	12.897	10.818	13.881	12.824	18.586
COE2	11.224	06.120	13.237	25.576	21.975
COE3	09.275	07.541	11.730	06.727	15.838
COE4	16.941	21.606	18.534	33.115	16.957

TABLE 14.6 COEs vs. AOPs Chi-Square Test P-Values

COE	Applicability of Online Education				
	AOP1	AOP2	AOP3	AOP4	AOP5
COE1	0.377	0.545	0.608	0.686	0.291
COE2	0.510	0.910	0.655	0.060	0.144
COE3	0.679	0.822	0.762	0.978	0.464
COE4	0.152	0.050*	0.294	0.007*	0.388

*Statistically significant.

- **H_{03}: The challenges of accessing digital technology are not associated with lack of awareness (COE11), anxiety or fear of technology (COE12), lack of technical exposure (COE13), or improper communication practices (COE14).**

The test statistics prove that lacking awareness (COE11) is not a significant concern that causes a problem in accessing digital technology ($p = 0.028 < 0.050$). However, anxiety or fear of technology (COE12), lacking technical exposure (COE13), and improper communication practices are significantly contributing to it (Table 14.7).

TABLE 14.7 COEs vs. Accessing Online Technology

Technological Challenges	Chi-Square Value	P-Value
COE11	28.381	0.028*
COE12	25.753	0.058
COE13	19.882	0.226
COE14	16.428	0.423

*Statistically significant.

- **H_{04}: The challenges of accessing online education content are not associated with technology handling (COE21), low motivation (COE22), lack of comfortability (COE23), lack of in-person experience (COE24), lacking an individual approach (COE25), and feeling of loneliness (COE26).**

The test results conclude that all the factors under study, i.e., handling techniques (COE21), low motivation (COE22), decreased comfort (COE23), lack of personal experience (COE24), personal attitude (COE25), and feeling of loneliness (COE26), contribute significantly to the problem of accessing online education content (all p-values are found to be greater than 0.05; Table 14.8).

TABLE 14.8 COEs vs. Content Accessibility Challenges

Content Accessibility Challenges	Chi-Square Value	P-Value
COE21	20.561	0.057
COE22	7.044	0.855
COE23	18.551	0.100
COE24	18.561	0.100
COE25	18.425	0.103
COE26	13.464	0.366

- **H_{05}: Resistance to change (COE31), lack of connectivity with peer groups (COE32), and inability to establish virtual connections (COE33) are not responsible for users' inconvenient engagement with online education.**

It has been found that all three factors – resistance to change (COE31), lack of connectivity with peer groups (COE32), and inability to establish virtual connections (COE33) – are significantly associated with users' discomfort in OLL (Table 14.9; all the p-values are found to be more than 0.05).

TABLE 14.9 COEs vs. Comfortability Associated Challenges

Comfortability-Associated Challenges	Chi-Square Value	P-Value
COE31	15.948	0.194
COE32	10.598	0.564
COE33	13.552	0.330

- H_{06}: In accepting online education, the non-availability of equipment (COE41), lack of technical proficiency (COE42), lack of online learning skills (COE43), and inability to engage with virtual students (COE44) do not cause any instructor challenges.

It has been found that the lack of technical proficiency (COE42) has a significant impact on instructors' training-related challenges (p=0.051>0.05). Users do not perceive other factors, such as the non-availability of equipment (COE41), lack of online learning skills (COE43), and inability to engage with virtual students (COE44), as associated challenges in accepting online education (Table 14.10).

TABLE 14.10 COEs vs. Trainer Associated Challenges

Trainer-Associated Challenges	Chi-Square Value	P-Value
COE41	24.947	0.015*
COE42	20.991	0.051
COE43	30.677	0.002*
COE44	23.966	0.021*

**Statistically significant.*

14.5 RESEARCH IMPLICATION

Online education is considered as the only option to educate the public in this pandemic situation. Governments are advising their educational institutes to provide online education to maintain social distancing without compromising the education services. This study identifies the challenges of present online education practices in order to make it more effective. The findings of this study are particularly useful for governments of developing countries and their educational institutes. It identifies the priorities of various users of online education, which can help governments and educational institutes frame more applicable policies for online education. Undoubtedly, all governments and educational institutes are considering online education to educate the public during a pandemic situation. However, there are still some associated challenges with its present format and practices that limit its acceptance and effectiveness.

The advanced education institutes offering advanced certification programs need to work more on sharing effective learning. Advanced certificate aspirants are facing challenges in accessing technology, not finding desired comfortability with the platform, and not finding qualified instructors. Additionally, the study suggests that OLL platforms should be more interactive and user-friendly. Most users face these two challenges. As shown in Table 14.10, users face content accessibility challenges due to their inability to deal with technology (COE21), lack of convenience (COE23), lack of personalization experience (COE24), and lack of desired customization (COE25). This report highlights that college-going students and parents are the most affected segments by COE services. Education institutes need to establish better connectivity with them to enhance acceptance and effective learning. Present platforms used in online education are either too ordinary or too complex to use. These platforms should be easy to understand and use, minimizing usage-skills, content-usage, and technology-accessing related challenges (Figure 14.4).

14.6 CONCLUSION

Many research studies, experts, and articles are emerging in favor of online education. This research work highlights the challenges that currently exist in online education practices. The study evaluates the readiness of

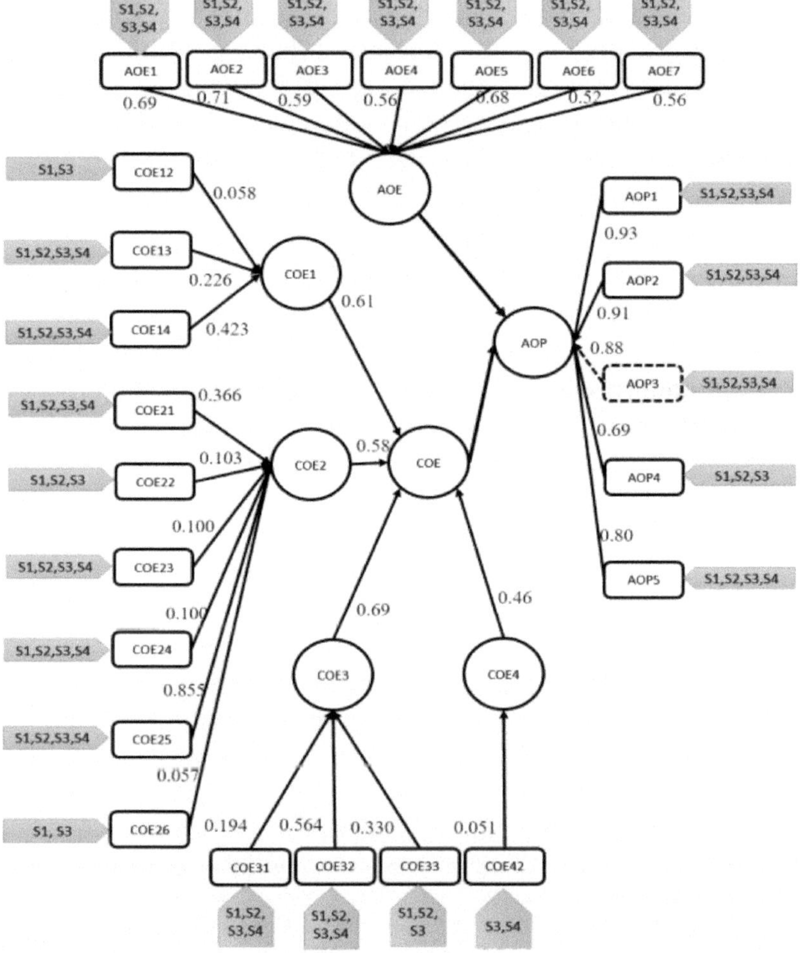

FIGURE 14.4 Factors and challenges associated with online education in developing countries.

developing countries, particularly India, to adopt online education platforms for the masses. There is a significant need to identify and improve the areas of concern in order to provide better online education in developing countries and take appropriate corrective actions. This study can contribute significantly to enhancing the effectiveness and AOE, especially during times of epidemics, in educating people. The study found that the lack of digital technology (COE1) and limited experience with current formats and practices (COE3) are hindering its acceptance and scope. Although the study delved deep into each considered challenge, further research can be conducted to identify innovative ways to overcome these challenges or minimize their impact. Moreover, it is important to note that this study is based on a small sample of 85 users. A more extensive study with a larger sample size may provide better insights.

KEYWORDS

- **community education**
- **COVID-19**
- **critical success factors**
- **developing economy**
- **educational institutes**
- **online education**
- **pandemic situation management**

REFERENCES

Atiaja, L. A., & Proenza, R., (2016). The MOOCs: Origin, characterization, principal problems, and challenges in higher education. *Journal of e-Learning and Knowledge Society, 12*(1).

Bostan, S., Erdem, R., Öztürk, Y. E., Kılıç, T., & Yılmaz, A., (2020). The effect of COVID-19 pandemic on the turkish society. *Electronic Journal of General Medicine, 17*(6).

Brandt, J. O., Barth, M., Merritt, E., & Hale, A., (2020). A matter of connection: The 4 Cs of learning in pre-service teacher education for sustainability. *Journal of Cleaner Production, 279*, 123749.

Cheawjindakarn, B., Suwannatthachote, P., & Theeraroungchaisri, A., (2013). Critical success factors for online distance learning in higher education: A review of the literature. *Creative Education, 3*(8), 61.

Conrad, D. L., (2002). Engagement, excitement, anxiety, and fear: Learners' experiences of starting an online course. *The American Journal of Distance Education, 16*(4), 205–226.

Esfandiari, Z., Mirlohi, M., Tanha, J. M., Hadian, M., Mossavi, S. I., Ansariyan, A., & Pakmehr, F., (2020). Effect of face-to-face education on knowledge, attitudes, and practices toward "traffic light" food labeling in Isfahan society, Iran. *International Quarterly of Community Health Education*, 0272684X20916612.

Gillett-Swan, J., (2017). The challenges of online learning: Supporting and engaging the isolated learner. *Journal of Learning Design, 10*(1), 20–30.

Global Online Education Market- Forecasts From 2020 To 2025. https://www.researchandmarkets.com/reports/4986759/global-online-education-market-forecasts-from (accessed on 20 February 2024).

Gourley, B., & Lane, A., (2009). Re-invigorating openness at the open university: The role of open educational resources. *Open Learning, 24*(1), 57–65

Hair, Jr. J. F., Sarstedt, M., Hopkins, L., & Kuppelwieser, V. G., (2014). Partial least squares structural equation modeling (PLS-SEM). *European Business Review.*

https://www.technavio.com/report/online-education-market-in-india-analysis-share-2018 (accessed on 20 February 2024).

https://www.theguardian.com/education/2020/mar/17/universities-having-to-adapt-fast-to-the-coronavirus-crisis (accessed on 20 February 2024).

Jokivirta, L., (2006). *What Went Wrong with All Learn?* University Business.

Kelley, E., & Hurst, J., (2006). *Health Care Quality Indicators Project.*

Lewnard, J. A., & Lo, N. C., (2020). Scientific and ethical basis for social-distancing interventions against COVID-19. *The Lancet Infectious Diseases, 20*(6), 631.

Long, N. J., (2020). From social distancing to social containment: Reimagining sociality for the coronavirus pandemic. *Medicine Anthropology Theory.*

Martínez-Martínez, A., Cegarra-Navarro, J. G., Garcia-Perez, A., & Wensley, A., (2019). Knowledge agents as drivers of environmental sustainability and business performance in the hospitality sector. *Tourism Management, 70*, 381–389.

Mashile, E. O., & Pretorius, F. J., (2003). Challenges of online education in a developing country: Research in higher education. *South African Journal of Higher Education, 17*(1), 132–139.

Narjis, G., & Shabbir, J., (2018). Estimation of population proportion and sensitivity level using optional unrelated question randomized response techniques. *Communications in Statistics-Simulation and Computation*, 1–15.

Ridenour, C., Simmons, A. L., Ilg, T. J., & Place, A. W., (2005). *'Divertual' Learning in Education Leadership: Implications of Teaching Cultural Diversity Online vs. Face to Face.*

Selim, H. M., (2007). Critical success factors for e-learning acceptance: Confirmatory factor models. *Computers & Education, 49*(2), 396–413.

Terpstra, J. T., (2004). On estimating a population proportion via ranked set sampling. *Biometrical Journal: Journal of Mathematical Methods in Biosciences, 46*(2), 264–272.

Tesar, M., (2020). *Towards a Post-COVID-19 'New Normality?': Physical and Social Distancing, the Move to Online and Higher Education.*

Volery, T., & Lord, D., (2000). Critical success factors in online education. *International Journal of Educational Management.*

Zamanzade, E., & Wang, X., (2017). Estimation of population proportion for judgment post-stratification. *Computational Statistics & Data Analysis, 112*, 257–269.

CHAPTER 15

Antecedents of the Crisis in Indian Management Education, and the Role of an Interdisciplinary Curriculum in Overcoming the Crisis

KIRTI PRASHAR,[1] TANVEER KAUR,[2] SIMERJEET SINGH BAWA,[3] and MINAKSHI SHARMA[4]

[1]*Assistant Professor, Sharda University, Greater Noida, UP, India*

[2]*Research Scholar, University School of Business, Chandigarh University, India*

[3]*Associate Professor, Chandigarh Group of Colleges, Landran, India*

[4]*Research Scholar, Master of Public Health, University of North Florida, Jacksonville, Florida, United States*

ABSTRACT

The idea of management is not a new one; it has existed for as long as there have been people. To achieve desired outcomes, every idea and life event must be managed. Similarly, management is essential for the functioning and structure of any organization. The strength of competent management is the cornerstone of any successful firm. In simple terms, management is about doing the right thing in the right way to accomplish stated goals. Timely goal completion is facilitated by effective management, which ensures the efficient and effective use of resources with minimal waste. The objective of this research chapter is to examine the

Interdisciplinary Approaches in Management Education.
Tanusree Chakraborty, Ramandeep Saini, Rajiv Khosla, Nandita Mishra, and Ashok Natarajan (Eds.)

concept of management. It helps readers understand how management relates to other disciplines. The study chapter sheds light on issues in management education (ME) and the challenges faced by managers. The study chapter has significant implications as it provides insights into the future of management in India. The chapter's study will guide students in understanding management studies and its scope, enabling them to grasp the practical approach of the subject.

15.1 INTRODUCTION

In order to achieve an organization's ultimate aim, management must effectively administer, coordinate, and put into practice all policies, strategies, and procedures. Since people are a company's most valuable asset, human resources are most often associated with management. All management levels, including the top, middle, and lower levels, must be properly coordinated and synchronized.

- **Management as Art and Science:** Both art and science are used to describe management. In a sense, management is a science, and like science, it has tenets that are widely recognized. It also investigates the causal relationships between different choices and outcomes. However, management is sometimes referred to as art since, like art, it requires practical knowledge, practice, individual talent, creativity, etc. In order to make the greatest management judgments.

15.1.1 HIGHLIGHTS OF THE CHAPTER

The chapter stresses the limitations of ME in India and highlights recent trends that have been adopted to improve it. The chapter will shed light on potential solutions to enhance the efficiency of ME.

15.1.2 OBJECTIVES

- To learn the fundamentals of management;
- To educate managers about current trends and difficulties;

- To research how management relates to other disciplines;
- To shed light on the management of the future.

15.2 RESEARCH METHODOLOGY

The exploratory study research methodology is employed in the study because it aims to analyze management-related themes in order to identify problems and obstacles. The study chapter solely relies on secondary data and a review of the literature. The purpose of the chapter is to shed light on the potential of management by examining the problems and difficulties faced by management and assisting the manager in understanding how to turn these difficulties into advantages.

15.3 LITERATURE REVIEW

Sanjeev Kumar et al. (2011) conducted a study on ME, which revealed the need for customization and integration of ME to bridge the gap between academia and industry requirements. The study emphasized the importance of grooming students, creating corporate awareness, fostering attitude change, and developing managerial skills. In a study by Sridhar (2015), it was stated that many management programs, including postgraduate programs, still fail to provide practical exposure to students, which is crucial in the current scenario. These issues should be addressed in the future. Rao et al. (2014) used a similar framework by surveying employers to determine what they look for when recruiting MBA graduates. The study also found that while top-tier institutes focus well on the knowledge dimension, there is still room for improvement in areas such as creativity, ethics, commitment to the organization, and learning attitudes, which fall under the doing and being dimension. Kapur & Perry (2015) studied the rapid growth of higher education (HE) in China and India. They felt that the mere growth in the number of students or expenditure in HE does not justify the meaningfulness of creating such infrastructure. They have clearly highlighted the research gap by concluding that researchers have found it extremely difficult to assess two critical outcome measures: first, the quality of HE beyond research publications, and second, the value added by HE over students' innate abilities.

According to Salunkhe & Pankajashan (2015), power, urbanization, town planning, renewable energy resources, rural areas, and water have a very close interface and interaction with each other. Therefore, there is a need to retool the educational philosophy, structure, and curriculum of all professional educational streams. Functional expertise will have to be juxtaposed with ME to deliver value in addressing the integrated challenges of the business and the changing business environment.

In 2020, Archie authored a research chapter about management problems. The chapter explored three main subjects: current research, the future potential of the field, and the advantages and disadvantages of management techniques vs. emerging research fields. To determine the most relevant choice for management methods, a study was conducted using the questionnaire method. Knowledgeable managers assisted the researcher in identifying the top management techniques. In their study, Vojko & Zlatko (2021) examined the utilization of different management strategies in firms. The main objective of the research was to align management techniques with those of industrialized nations. To investigate management practices in well-organized industrialized nations, data was collected from a sample size of 155 respondents. The study identified three key management practices for ensuring organizational development and survival: customer relationship management (CRM), enhancing customer interactions, and promoting customer satisfaction. The report also suggested that in order for firms in developing nations to compete with those in wealthy countries, they need to adopt the same processes and strive to establish strong customer relationships.

15.4 MANAGEMENT EDUCATION (ME) IN INDIA

The business schools of Iverson (1998) have stated that ME in India is still in the experimental stage. It is in its early stages, an invention of the 20th century. Business education in India needs to be customized in the context of institutional frameworks and the socio-cultural environment, which seems to be the new trend among business schools in developing countries in the coming years. In India, business management is formed by a variety of elements, including the knowledge, skills, requirements, and preferences of business education consumers, as well as the abilities required by employers in management graduates. The backdrop for ME essentially comprises each of these parts. The educational context is also influenced by socioeconomic factors (Bush, 2007; Bush et al., 2010). The personality of any business school is shaped by its cultural and institutional settings (Bush et al., 2010). The Indian setting today differs from what it was a few years ago and from what it was in the Western world.

According to the All-India Council of Technical Education's (AICTE) Annual Report for 2012, there are a total of 3,849 AICTE-accredited management colleges that offer management degrees, as well as 327 unapproved institutes that offer management degrees. However, India has still failed to develop the proper curriculum for management instruction. Many leading management institutions in India use the same curriculum and books as those used in the Western world, without even analyzing the relevance of the curriculum in the context of the Indian environment.

15.4.1 AILS IN PRESENT MANAGEMENT EDUCATION (ME)

Management research and education are criticized for focusing on scientific and academic rigor at the cost of the relevance of organizational understanding (Bennis & O'Toole, 2005). Theories of bad management point to Ghosal (2005), damaging the best practices of management. The curriculum of business management courses tends to prioritize the concerns of shareholders over stakeholders, which helps guide students in their roles as managers and leaders to adopt moral responsibility (Ghosal, 2005). Chia & Holt (2008) argue that curricula should prioritize content that can be classified as "knowledge-by-representation" rather than "knowledge-by-exemplification." MBA programs have been criticized for failing to adequately prepare corporate leaders. What academic researchers and others often forget is that business education is not the same as traditional academic fields. Professional schools are obligated and challenged to create both exoteric and esoteric knowledge that is relevant to business leaders. The primary focus of concern revolves around the MBA program. The results of large-scale surveys conducted by the Graduate Management Admission Council consistently reveal concerns about learning outcomes and curriculum. The challenges of the future are already becoming apparent. The majority of business schools are struggling to receive positive feedback from students. The number of applicants is decreasing, which is a clear indication that business schools need to brainstorm solutions to these problems (Trilok, 2019).

15.5 ENHANCEMENT IN AILS FOR MANAGEMENT STREAM AFTER COVID-19

In many respects, the COVID-19 catastrophe served as a wake-up call, reminding business executives and business schoolteachers of the significance of preparing for worst-case scenarios such as communicable diseases, political unrest, and monetary destabilization, among other things. Prospective MBA students should look for schools that can help them develop broader abilities such as innovation, systems thinking, and managing and assisting workers through change. While change happens more swiftly in times of crisis, it happens all the time in business (Ilana Kowarski, 2020). During the crisis, companies like Zoom, a video conferencing firm, and Slack, an office communications tool, saw their stock prices surge. To handle the rush for a vaccine and the global demand on hospitals and medical providers, pharmaceutical and healthcare businesses will need skilled, trained management personnel. While incomes, job offers, and acceptances did fall during the worldwide pandemic's health catastrophe, which started in March 2020, the "most substantial impact" was not in full-time employment recruiting at all. Instead, the number of internships that lead to jobs has plummeted (Standards for Reporting Full-Time MBA Employment Statistics).

15.5.1 RECENT TRENDS IN MANAGEMENT FOR OVERCOMING THE AILS DIVERSE WORKFORCE

Managers face challenges when managing a staff with diverse cultures, languages, and backgrounds. It can be difficult to coordinate a workforce with varying mental states. It is important for supervisors to have a proper understanding of the psychological state of their personnel. In addition to dealing with professionals from various professions and backgrounds, management also faces the issue of coordinating individual interests in order to achieve the overall goals of the company.

15.5.2 UPDATING EMPLOYEES WITH LATEST TECHNOLOGY

The age of invention is now. Technology is always evolving. If the staff doesn't adapt to the changing environment, the company won't be able to endure over time. It might be difficult for managers to provide frequent training to the staff so that they won't resist adapting to their surroundings. The company will benefit from regular upgrades as it makes it easier to identify opportunities early and reap rewards.

15.5.3 MANAGING THE STRESS

The world is full of stress and problems. The workplace atmosphere must be maintained so that staff members feel comfortable carrying out their jobs. Employees' work may suffer as a result of stress. To ensure that employees can provide their best to the companies and perform effectively, managers must frequently adopt appropriate stress management measures. It is necessary to establish suitable grievance cells and identify specific criteria in order to manage stress.

15.5.4 STIMULUS FROM DYNAMIC ENVIRONMENT

Every small change in the social context will undoubtedly have an impact on how the company operates, as it operates within that environment. Changes in culture, demographics, society, and geography all have an equal impact on business. For example, the traditional handloom industry in India is being influenced by Western culture. Changes in the business will follow the economy's gradual recovery. GDP, per capita income, fiscal policy, monetary policy, and other factors all have an effect on the economy.

The primary goal of management is to deliver high-quality results for the company. Management must ensure a conducive work environment and gather all necessary resources to maximize the company's profitability and efficiency.

15.5.5 HANDLING OF GRIEVANCES

The management must be careful when resolving complaints because of the many conflicts that exist between what is best for an individual and what is best for the company due to the ongoing changes in society. If the complaints are addressed well, the staff will feel satisfied, identify more with the company, and work harder to achieve the general objectives of the company.

15.5.6 COOPERATION WITH POLITICAL ISSUES

Politics has an ongoing effect on the economic world. There are several changes during elections, such as campaigning, etc. Several new amendments have been made to the code of business behavior as a result of changes in the government.

15.6 FUTURE OF MANAGEMENT

Corporate Culture: According to the holistic approach, management is the foundation of everything happening in the organization. Management is the guiding force for every corporate behavior. Corporate culture basically represents the rules, accepted norms, and models associated with that company.

15.6.1 EMPLOYEES

Employees are the first point of contact for every customer of the company. The behavior of the employees acts as a representation of the company's service and reputation. The employees are also considered as ambassadors of management ethics. How the employees are dressed, how they interact and behave with customers will leave an impression about the brand or the company in the minds of the customers. In the service sector, the importance of employees' behavior and body language in exhibiting values and acting as ambassadors.

15.6.2 COMMUNICATIONS

Management in the corporate world nowadays depends on the effective efforts of all functions of the company to communicate effectively with the target audience. According to Malmelin and Hakala in Radikaali Brandi, for companies nowadays, management is all about assimilating information and communication in such a way that all functions of the organization are in accordance with the needs of customers. The firm also communicates various intrinsic values to the target audience and stakeholders through visual elements. Due to this communicative characteristic, it is very important to understand and deliver accurate communication of the management.

15.6.3 LEADERSHIP

The management and leadership of any organization have an integral role to play through their actions. In today's time, the scope of management is not just limited to market-oriented phenomena, but it is a mindset adopted by the entire organization, from top-level management to bottom-level management. It is believed that the CEO of the organization is the person who should be responsible for making all decisions in today's times. Good and effective leaders of the company instill a lot of confidence in the investors because of their capabilities and competencies.

15.6.4 IMAGE AND VISUAL IDENTITY

The image and visual identity of a firm are among the most traditional elements of marketing. The designs and visual effects are used by the company to convey its values and beliefs through the use of graphics and colors. Various symbols, colors, etc., have different meanings for different audiences and are therefore used for communication in a very subliminal manner. The colors or images that a company uses can be considered an essential tool for creating a certain type of brand.

15.6.5 QUALITY OF EDUCATION

Management education (ME) is among the most coveted disciplines. Candidates seek quality education in management that can help them understand economies and business, as well as provide practical exposure. Corporate firms are looking for job-ready graduates from the management field who are able to work under pressure, are ethical, flexible, willing to learn, can recover from mistakes, are team players, can manage tasks and solve problems, inspire others, set goals, and are result-oriented. In Europe, employers focus on candidates who are managers and leaders.

15.6.6 DIVERSITY

The decade has seen significant changes in management schools. Diversity has been observed in the student profile, with students from various industries being invited back to classrooms. ME is becoming increasingly global and is considered a necessity in the multinational business environment. It is important for ME to also include knowledge of humanities and an appreciation of other cultures, considering the diversity found in the corporate world.

15.6.7 INDUSTRY INVOLVEMENT

Senior executives are becoming more conscious of their contribution towards education and are very willing to make time for involvement across B schools. They serve as mentors, guest speakers, visiting faculty, and collaborate on research projects, providing students with opportunities to learn in real-world scenarios under the guidance of experts. Two emerging trends can be observed – (i) management education is becoming globalized; and (ii) the use of technology is increasing.

15.6.8 CURRICULUM

The curricula for ME have been revised over time to incorporate practical models and practices. Trainings and internships have also been included as part of the curriculum, allowing students to engage in practical projects and gain knowledge about the real world. The emphasis has shifted from learning to implementation.

15.6.9 PEDAGOGY

Pedagogy generally includes the teaching process, the richness of the library, the number of courses offered, and co-curricular activities to enhance students' management skills. The general perception of ME is that it focuses on functional areas such as marketing, finance, and human resources, as well as a few courses like accounting and statistics. However, there is a need to have a holistic view and integrate the learning received from various courses.

15.6.10 GLOBAL EXPOSURE

The world has become smaller and business schools need to adapt. Students are now seeking more and more global exposure. Exchange programs are a part of the business school curriculum, and joint degrees have emerged as a new offering to diversify the learning experience of students. The traditional method of case-study is being enhanced by immersive global experiences, where students can engage in overseas live projects. By having this kind of world-level exposure in different settings and contexts, students can be prepared for leadership roles at a global level.

15.6.11 FACULTY SHORTAGE

In India, many B-schools do not think in a broader sense. Indian institutes do not have skilled teachers who can teach courses that are solid enough and bring real-world level experiences to the classroom. Additionally, Indian institutes have also failed to attract well-qualified practitioners from reputable companies to impart knowledge to the students. In every aspect, it can be said that every branch of learning requires in-depth technical and professionally-oriented critical thinking in this era of digitalization.

15.6.12 ONLINE MANAGEMENT PROGRAMS

The demand for online programs is continuously rising despite various drawbacks. These programs are provided by many institutions for budgetary reasons. These institutions must initially spend in order to provide online programs. The cost per student, however, is quite inexpensive. Since there is no cap on the number of students who can enroll, many institutions view these programs as potential sources of income. Due to low government funding, several famous colleges now offer online MBA and other programs. The goal is to open up HE to more students who are unable to enroll in traditional programs.

15.6.13 CHANGING MANAGEMENT EDUCATION (ME) SCENARIO

Across globe there is emerging changing scenario in the field of ME. With changing trends now business colleges of India are forced to work at international level business, which focusing on creating graduates with mindset fit for global requirements as well as in order to develop better links overseas. In order to improvise the ME quality, Indian education institutes have been looking for tie ups with many universities in abroad. Apart from this, aspirants in India prefer foreign institutes from UK, Australia, and Canada and so on for the degree of MBA.

15.6.14 AN ORGANIZATION

An organization can only be impacted by management. If the management is knowledgeable, that is, if the management is made up of professionals from various disciplines who have an in-depth understanding of the subject at hand, then it will provide a firm foundation on which the company will operate. The existence and development of the firm are initially reliant on the business idea, but over time they become reliant on the effectiveness of the management.

15.6.15 EDGE OVER COMPETITORS

Competition has a direct impact on a company's profitability and capacity to survive. The management of the business must examine the strategies and tactics utilized by its competition in order to gain an advantage over them, enabling the firm to align its own strategies accordingly. The management will continuously strive to create new strategies that surpass the competition in terms of pricing and quality. By consistently anticipating the right adjustments, a company can position itself ahead of rival companies.

To accurately evaluate and assess the actual amount of profit generated, the management must prioritize the creation of tangible results.

15.6.16 EFFECTIVENESS IN ASSESSING OPPORTUNITIES

Opportunities are always hidden assets for an organization, with the potential to increase revenues. The management should be able to assess prospects efficiently and turn them into successful ventures. Today, a company's success is based on how well its management can predict the future and take advantageous actions.

15.7 WHAT MORE CAN BE DONE IN FUTURE?: A WAY TO LOOK FORWARD

Research that focuses on industry connections is critical. According to most survey participants, there is a need for industry linkage, which can only be achieved through excellent research that addresses industry requirements. Strong research will provide teachers with the most up-to-date information and assist industries in enhancing their performance. Consequently, institutions should modify their roles and consider themselves as supporters of the industry.

15.7.1 PRACTICAL PROJECTS

Students in business schools should be assigned practical projects that are based on real-world business concerns. These initiatives will allow students to focus their studies on real-world issues and develop skills that are highly sought after in the corporate world. This will also provide students with a pragmatic approach to education. Additionally, students will be better prepared to face challenges in the real world.

15.7.2 TECHNOLOGY INVESTMENT IS CRITICAL

Technology is evolving rapidly, surpassing expectations in terms of the pace of change. Today's business schools focus on outdated technologies and philosophies, rather than teaching about current and future technology. The majority of library books are now outdated, as software and data have evolved. Therefore, it is crucial to prioritize the most up-to-date technologies and procedures. Outdated books do not benefit institutions, as they need to familiarize themselves with current company data and prepare students for the tools and approaches that will be utilized in the next decade.

Most business schools have failed to stimulate their students' minds through discussions and brainstorming sessions, due to a closed teaching approach. It is necessary to encourage dialogue and brainstorming sessions, allowing students to explore all their options. This will enhance students' performance and better equip them for the future. Providing ME in open classrooms, where professors facilitate discussions and classrooms transform into open debate forums, enables students to discuss, solve, and apply principles to real-world situations.

15.7.3 STARTUP EXPOSURE

Due to a lack of proper industry-institute connections, many students nowadays do not have the opportunity to gain exposure to the business sector. There is an increasing expectation among students to have a better understanding of the start-up scenario. The majority of start-ups are in the information, communication, and internet-based business sectors, as well as other related fields. There is a need for a strong link between business schools and start-up companies. This will enable students to create their own businesses. Students aspire to work in start-ups, but they require a supportive ecosystem within business schools, which can only be achieved if the institution's activities are properly connected.

15.8 RELEVANCE OF MANAGEMENT IN MANY OTHER BRANCHES: A REASON FOR MANAGEMENT TO UPGRADE

15.8.1 ENGINEERING AND MANAGEMENT

Essentially, engineering is a field that develops innovative techniques for streamlining processes. Engineering is ineffective if tools, equipment, and resources are not managed properly. Therefore, management and engineering are closely related. The right planning, organization, process, and tactics are used, much like in management. A superior engineering outcome is guaranteed by the correct channelization of all resources, which is only made feasible by effective management and testing for deficiencies.

15.8.2 HOME SCIENCE AND MANAGEMENT

Home science involves the practice of performing daily tasks at home, encompassing various aspects of leading a good life and carrying out activities such as earning a living and eating. Management plays a crucial role in home science, ensuring the smooth execution of daily activities.

15.8.3 ADMINISTRATION AND HEALTHCARE

While patient care, service delivery, and human skills are all integral to healthcare, patient care itself is not considered a function. Management is closely intertwined with the healthcare industry, as effective management is essential for the development of a proper healthcare sector.

15.8.4 FINANCE AND MANAGEMENT

Finance is the foundation of any business; it is not only a resource. Both a lack of capital and excessive funding can be harmful to the company's expansion. Therefore, management is essential to the efficient handling of money from the point of collection to the point of usage.

15.8.5 MANAGEMENT AND OPERATIONS

Manufacturing, sourcing, assembling, selling, warehousing, transportation, data management, and other activities are all included in operations. Management may aid in achieving the overall objectives of the company if all operations are well coordinated.

15.9 WHAT DO RECRUITERS WANT FROM MANAGEMENT EDUCATION (ME) GRADUATES?

There are many indicators that reveal business schools are slowly losing their importance, as graduates in this field aren't yet ready for the world of industry and business. Thus, institutions need to analyze the qualities that recruiters are looking for in students and should focus on developing those skills by combining theoretical education with practical exposure.

The most recent GMAC Recruiter Survey has provided the following list of qualities that recruiters seek in each business school graduate:

- Integrity;
- Professionalism;
- Adaptability;
- Initiative;
- Motivation;
- Innovation and creativity;
- Collaboration.

Employers in Asia and the United States prefer individuals who can perform well under pressure, demonstrate ethical behavior, adapt easily, have a willingness to learn, can recover from mistakes, work effectively in teams, manage projects, and solve problems, motivate others, set goals, and are results-oriented. In Europe, employers highly value individuals with managerial and leadership skills.

Employers are seeking applicants who can effectively organize and integrate information from multiple sources and apply their expertise to solve real-world problems and make predictions about the future. All recruiters emphasized the importance of high-level integrative thinking skills to tackle challenging tasks, such as merging and processing diverse data. Additionally, employers are actively seeking individuals with technical and analytical abilities.

Employers thus look for graduates who are a complete package, possessing the abilities to deal with the practical world and able to:

- Organize information to solve multiple problems;
- Combine, manipulate, and manage data;
- Synthesize information;
- Make judgments on information from various sources.

15.10 KEY TAKE AWAYS FROM THE CHAPTER

Today's ME needs to be transformed. There is a need to lay out a clear route for how to move forward and in which direction to go. The majority of business schools are preparing for tomorrow's difficulties. The current study is an attempt to bring to light towards the suggestions and issues that will affect the future. These problems will aid policymakers in developing a future strategy.

15.11 CONCLUSION

Human resources are integral to effective organizational management, serving as the company's most valuable asset and driving its operations. Proper coordination and synchronization across all management levels, including top, middle, and lower tiers, is essential. The concept of management is not new; it has existed since the dawn of civilization. Every idea and life event must be managed to achieve desired outcomes. Management plays a critical role in the operation and structure of any organization, combining elements of both art and science. It is a science, guided by widely recognized principles and exploring causal relationships between choices and outcomes. Yet, it is also considered an art, requiring practical knowledge, experience, individual talent, and creativity. Effective managers possess a deep understanding of both the art and science of management to make informed judgments.

The strength of effective management is the cornerstone of a successful organization. Management entails executing the right actions in the right manner to achieve stated objectives. Timely goal attainment is facilitated by efficient and optimal resource utilization. The featured study chapter holds significant implications for the future of management in India. ME is a highly sought-after field. Candidates seek quality ME to understand economies, business, and gain practical exposure. Corporate firms seek job-ready graduates who can work under pressure, are ethical, flexible, willing to learn, can recover from mistakes, are team players, can manage tasks and solve problems, inspire others, set goals, and are result-oriented. In Europe, employers prioritize candidates who possess managerial and leadership skills.

India has yet to develop a suitable curriculum for ME. Many leading management institutions in India adopt the same curriculum and books as those used in the Western world, without considering their relevance in the Indian context. This chapter highlights the need to revitalize ME in India. For a long time, ME in India has been criticized for being out of touch with current global practices. The chapter identifies the following areas for improvement in ME:

- Focus of management education;
- Method of delivery;
- Course redesigning as per the Indian environment.

KEYWORDS

- **dynamic environment**
- **home science**
- **management education**
- **modern issues**
- **online management programs**
- **stress management**
- **technology investment**

REFERENCES

AACSB International, (2020). *Final Report of the AACSB International Impact of Research Task Force. 24*(1), 123–129.

Bennis, W., et al., (2005). *How Business Schools Lost Their Way. 13*(4), 54–63.

Bush, T., & Glover, D., (2016). School leadership and management in South Africa. *International Journal of Educational Management, 30,* 211–231. 10.1108/IJEM-07-2014-0101.

Carroll, A. B., Adler, N. J., Mintzberg, H., Cooren, F., Suddaby, R., Freeman, R. E., & Laasch, O., (2020). What are responsible management? A conceptual potluck. In: *Research Handbook of Responsible Management.* Edward Elgar Publishing.

Chia, R., & Holt, R., (2008). On managerial knowledge. *Management Learning, 39*(2), 141–158. https://doi.org/10.1177/1350507607087579

Ghoshal, S., (2005). Bad management theories are destroying good management practice. *Academy of Management Learning & Education, 4,* 75–91. 10.5465/AMLE.2005.16132558.

Ilana, K., (2020). *The Relevance of an MBA During the COVID-19 Crisis.*

Jain, T., (2019). Transforming management education through Indian philosophy: Issues and pathways. *SSRN Electronic Journal.* 10.2139/ssrn.3318220.

Kapur, D., & Elizbeth, J. P., (2015). *Higher Education Reform in China and India, The Role of the State.* Havard – Yenching Institute Working Papers.

Kumar, S., et al., (2011). *Customer Relationship Management. 3*(1), 109–114.

National Knowledge Commission. (2019). *Report of the Working Group on Management Education.*

Nedelko, Z., (2021). What drives the usage of management tools supporting Industry 4.0 in organizations? *Sensors, 21*(10), 3512.

Prashar, K. (2020). *Empirical Study of Coronavirus and Its Impact on Economy. 2*(5), 1–6.

Prashar, K., & Sharma, A., (2019). Contemporary issues in management. *Think India Journal, 22*(27), 117–120.

Rao, T. V., (2014). *What Employers Value in the MBAs They Recruit: Rebalancing the Management Education Curriculum*. IIMA.

Salunkhe & Pankajashan, (2015). *Importance of Management Studies in Today's Competitive Environment*.

Sridar, K., (2015). *Past, Present and Future of Management Education in India*. *1*(9), 143–148.

Websites

http://www.financialexpress.com/jobs/shaping-the-future-of-management-education/65011/ (accessed on 20 February 2024).

https://www.hindustantimes.com/education/the-future-of-management-students-in-an-artificial-intelligence-based-world/story-SwyaGNdkNfhQ0AwmoujXWO.html (accessed on 20 February 2024).

https://indiaeducation.net/resources/interviews/ (accessed on 20 February 2024).

https://www.greatlakes.edu.in/chennai/future-of-management-education (accessed on 20 February 2024).

https://www.mbacsea.org/standards (accessed on 20 February 2024).

CHAPTER 16

Collaboration: A Key to Strengthening Industry-Academia Linkages

R. K. MISHRA[1] and SHULAGNA SARKAR[2]

[1]*Senior Professor, Institute of Public Enterprise, Hyderabad, Telangana, India*

[2]*DCM/HR, Learning and Development, NLC India Limited, Neyveli, Tamil Nadu, India*

ABSTRACT

Enhancing the quality of higher education (HE) has been an inspiring program for the government. Higher educational institutions are implementing various mechanisms to ensure quality in education. Industry–academia linkages drive the enhancement of quality in HE, offering a wide range of benefits that can positively impact educational institutions, organizations, society, and the nation as a whole. Collaboration is one such mechanism that strengthens industry–academia linkages, where members of an inclusive learning community work together as equals to support enhanced learning. On the other hand, industry–academia linkages, although not part of an inclusive learning community, also contribute to enhanced learning. This chapter aims to discuss the different forms of industry-academia collaborations and share their benefits. Additionally, the chapter addresses operational and technical challenges, along with strategies for institutions to strengthen industry–academia linkages. The chapter presents a case study of an academic institution that has actively

Interdisciplinary Approaches in Management Education.
Tanusree Chakraborty, Ramandeep Saini, Rajiv Khosla, Nandita Mishra, and Ashok Natarajan (Eds.)

practiced collaboration. It is a unique attempt that provides perspectives from both technical and management institutions.

16.1 INTRODUCTION

The society desires an education that builds character, expands mental horizons, enhances intelligence, and helps individuals develop their identity. In India, the Vedic sages pioneered the pursuit of higher knowledge since ancient times. In contrast, the existing educational system in India is only 140 years old, considering that the first universities were established by the British in 1857, after the First War of Indian Independence. Although the growth of higher education (HE) institutions in India was slow under British colonial rule, it accelerated after independence in 1947, with the establishment of several premier institutions. Since the early 1950s, the sector has experienced exponential growth, as evidenced by the increasing number of institutions, teachers, and students.

He is a key driver of a nation's development, which includes the government's policies to encourage enrollment in HE institutions, and policies to make HE more affordable to a larger mass of students. In recent times, the goal of sustainability has forced both universities and other HE institutions to rethink their curricula, research programs, and the employment potential of their graduates. Large-scale changes are being discussed and initiated for implementation. The challenges involved in the implementation of best practices are many and significant, and a successful transition for the HE system is full of possibilities.

The Indian education system has witnessed numerous reforms and transformations in the way it operates. Petrova & Sinclair (2008) have characterized the difference between traditional face-to-face educational systems and distance learning systems. They have shown that common interests, common pursuits, and a commonality of experiences, rather than the physical distance between the parties in a collaborative situation, help explain the success in cooperation and collaboration among institutions. Karamat & Petrova (2009) show that reducing communication costs further enhances concerted efforts.

Lai (2011) further points out that collaboration has a positive effect on performance, especially for low-achieving students. Although collaboration has been a concept more often applicable to individual participants, it also has high relevance at an organizational capacity. In educational institutions, collaboration is more of a coordinated and synchronous effort resulting in a sustained effort to build and uphold a unified approach to understanding a problem (Roschelle & Teasley, 1995). Vries et al. (2019); and Borrelli et al. (1995) share the advantages of collaboration while simultaneously emphasizing the difficulties in encouraging and establishing collaborations.

The chapter discusses different types of industry-academia collaborations and their benefits. It also covers operational and technical challenges, along with strategies to strengthen industry–academia linkages.

16.2 EDUCATION IN INDIA: THE PAST AND CURRENT SCENARIO

Education is a bridge that connects the past, present, and future, making it easier for future generations to learn from accumulated lessons of history and carry them forward. India has been a center for education and a practice-enriched culture for years. Teaching here is traditional in nature and dates back thousands of years. In India, we have witnessed the presence of great teachers like Chanakya, Vashisht, and Dronacharya. During the medieval period, India experienced a great synchronization of literature in both Indian and Islamic traditions, encompassing all fields of knowledge, science, arts, and sociology. Geographically, India, besides being one among the southern countries, is considered to have made a remarkable impact on Southeast Asian countries as well. The reawakening of India in the late 19th century has brought creativity, innovation, rejuvenation, and reconstruction to the country's education system.

HE, referring to post-higher secondary education, existed even in Vedic times, with Indian Rishis and sages providing it in the form of the pursuit of higher knowledge. We can call this HE as it represents the topmost layer of formal education. More importantly, it encompasses the processes of the most advanced stages of learning and development in the search for eternal truths and wisdom.

In contrast, the modern education system in India is only about 140 years old, with the first universities in India being established in 1857, following the end of the First War of Indian Independence. The British colonialists established the scope and role of HE institutions based on the policy guidelines provided by Macauley and Wood in their Despatch

(1854). Initially, Indian colleges were affiliated with British universities, but after 1857, their affiliation was changed to the newly-established universities. The development of progress in HE was slow in the intervening years up to independence.

16.3 GOVERNANCE AND MANAGEMENT

The role of higher education (HE) has become increasingly popular over the years as its importance is realized, accompanied by a growing aspiration among the population for HE. This has been accompanied by an increase in the accessibility and affordability of HE for the masses. HE institutions have responded by transforming themselves in terms of processes, revising curricula, and adopting technology-enabled delivery methods to attract students from all walks of life.

During this period of growth, HE institutions have faced several challenges. These include fostering multi-disciplinary collaborations, implementing technology-enabled learning, retaining faculty members, and catering to a diverse clientele. Additionally, there has been a proliferation of institutions beyond universities, providing a booming opportunity to utilize digital platforms and distance learning initiatives.

The governance of HE institutions has also undergone changes, with the involvement of various stakeholders in decision-making processes for the identification and implementation of new initiatives. This implies that institutions now seek more autonomy, involving all stakeholders in shared governance, while addressing challenges such as faculty recruitment and retention, establishing excellence standards, and designing curriculum. At the same time, institutions remain accountable to the system.

The HE system has been experimenting with various management approaches to tackle challenges arising from factors such as changes in academic disciplines, adoption of new instructional methods, and peripheral factors like population growth, diverse clientele, and evolving market requirements. Non-university institutions and the establishment of alternative educational institutions like open universities and distance learning systems have made significant progress and are impacting the educational system. Collaboration advocates emphasize the need for universities to engage in multidisciplinary and interdisciplinary collaborations with other entities, making them more socially, politically, and economically relevant (Schneidewind et al., 2009; Reale & Primeri, 2015).

16.4 COLLABORATION: THE CONCEPTUAL MODEL

The role of a collaborative approach in HE and systems, and the implementation of various positive outcomes such as critical thinking, improved meta-cognition, as well as the encouragement of underprivileged individuals, have been well established. Collaboration serves as a means to enable shared meaning and includes the feature of interactions among participants, allowing them to converge on shared meaning through the construction, monitoring, and repairing of shared knowledge (Roschelle, 1992).

Several studies have examined successful university-industry collaboration (Sandberg et al., 2011; Prigge, 2005; Calder, 2007). Cyert & Goodman (1997) have made a major contribution in this regard. They explore how problems are selected, processes are managed, teamwork is enhanced, and information dissemination occurs to develop synergy between universities and the corporate world. Similarly, Awasthy et al. (2020) have conducted extensive work in designing frameworks to improve university-industry collaboration. It is no wonder now that the ability to work collaboratively with others is seen as an essential skill in fulfilling this objective.

16.5 QUALITATIVE ANALYSIS

An attempt was made to approach various higher educational institutions from different countries. Responses were received from 11 Indian HEs, five HEs from Bangladesh, one HE from Bhutan, one from Nepal, and one from Pakistan. Data was collected through personal connections and convenience in sharing information. Interviews were conducted with academic leaders and practitioners. The practices shared by the interviewees were summarized in the form of a Collaboration model as shown in Figure 16.1. The details of the shared information were further explained in various forms of collaboration identified.

FIGURE 16.1 The collaboration models.

Source: By authors.

The conducted interviews included open-ended questions and were designed with a focus on discussion. This approach allowed for an understanding of how collaboration is facilitated in the institutions. The discussions revealed more individual perspectives than what was practiced in their respective institutions. The perspectives of the interviewees were summarized in a model, which clearly indicated a strong inter-linkage between various forms of collaboration. This is reflected in Figure 16.1. The data further clarifies the benefits of each form of collaboration, as presented in Table 16.1. The discussion also touched upon the practices of collaboration on papers with low or no impact. Enabling sustainability in collaboration is highly challenging, and the role of leadership is vital in driving the impact and sustainability of collaborative efforts.

TABLE 16.1 Qualitative Data Analysis

Academic Collaboration	Corporate	Research	Teaching
(a) Industry–university Branding Funding (b) Institute–institute Stronger research collaboration Better teaching Reputation Exposure to students	Placements Corporate consultancy Establishing corporate chair Increased funding	Sectoral research Need-based research Stronger research	Industry needs-based teaching Industry ready students Encouraging collaborative learning

Source: By authors.

16.6 INDUSTRY-UNIVERSITY LINKAGE

He is increasingly driven by the twin factors of collaboration and innovation. Furthermore, global agendas such as socio-economic growth, globalization, climate change, as well as emerging trends of big data and digitalization, are being discussed more and more in the public sphere and require increased awareness and education by all. The privileged position of research held by universities and HE institutions has been challenged by changes in business operations within the industry. This has led to collaboration playing a vital role in the traditional drivers of research, consultancy, and training for HE institutions.

The framework laid down by the Ministry of Science and Industry of the Indian government (Government of India, 2019) for the industry–university linkage in research reflects that despite the efforts made by the industry and academia to engage in collaboration, there have not been noticeable results observed so far. This has led to the conclusion of an absence of a strong and clear policy in this regard (Rangarajan, 2021).

Ghosh (2019) documents that the linkage between industry and academia in India is rated only 4.7 out of 10. This work also shows that the weak innovation ecosystem in academia has led to restrictions in a few areas dominantly. These include the capacity building in research, length and breadth of innovation, and a lesser number of enrollments in doctoral programs. It can be found in this study that out of 500 industrial organizations in India, 30%–35% do not have any research institute within the campus or vicinity that can foster research.

16.6.1 INSTITUTE–INSTITUTE COLLABORATION

It has been recognized that an integral part of learning is encouraging classroom activity, and interdependency among students forms a major component of collaboration. Similarly, the networking skills as well as the interpersonal skills of students are enhanced by encouraging collaboration among inter-institutional students. Collaboration in HE for India and Australia includes various types of collaboration in recent times. Looking at the international collaborations more specifically, it can be asserted that Australia and India have an augmented level of collaboration, with 134 new agreements signed between the institutions of these countries in recent times. The areas of these collaborations include franchising, exchange programs, twinning, and articulation (Table 16.2).

TABLE 16.2 Forms of International Collaboration

Franchising	Twinning	Program Articulation	Student Exchange
• Australian institutions appoint institutions in India as franchisees, authorized to provide all or part of their courses in India, towards a qualification awarded by the Australian institution.	• Under these arrangements, Indian institutions deliver courses specified and developed by Australian institutions. Students undertake the early parts of their studies in India, before completing the requirements of an Australian qualification in Australia. This reduces the cost and length of students' stay overseas and students benefit with an international exposure.	• Students can study the courses of the first few years at the Indian institution, with the Australian institution recognizing the credits they earn as advanced standing in specified courses.	• In these arrangements, the students complete a semester or two, or a project, in an institution overseas by prior arrangement, which earn credits towards their qualification awards. In some cases, the exchange is mutual, i.e., students from both institutions travel; in others, it is one way.

Source: By authors.

16.6.2 ENTRY OF FOREIGN UNIVERSITIES

The advent of globalization has led to universities and HE institutions expanding their reach beyond their own geographical and political boundaries. Renowned British, American, and Australian universities have established campuses in numerous other countries. However, their focus has primarily been on providing certification programs rather than direct teaching. As a result, they have formed collaborations with other HE institutions or even franchisees who deliver courses under the umbrella of the prestigious brand, while the universities themselves are not directly involved in teaching or imparting education.

16.6.3 CORPORATE COLLABORATION

16.6.3.1 EMPLOYABILITY ENHANCEMENT

In India, 0.8 out of 1.3 billion people are of working age. However, research by CII and its partners reveals a significant gap between the quality of skills available and the skills required. Approximately three-quarters of businesses in India face the challenge of a lack or shortage of technical or specific skills among their workforce. Additionally, nearly two-thirds of businesses struggle to find employees with overall employability skills, including effective communication, teambuilding, and other behavioral competencies. Around 61% of businesses also face challenges related to the adequacy of job applicants' applications and work experience. Research indicates that the talent and skills of the workforce tend to decline as they age. It is crucial for academia and governments to collaborate in providing the necessary skills training to the younger generation, leading to higher levels and quality of employment. Collaboration among academia, students, employers, and the government is essential to address the significant talent gap in India.

Haugh & Talwar (2010) conducted a research chapter to examine sustainability in well-known large Indian companies. The chapter focuses on implementing sustainability in major corporations through a three-tier learning module that encompasses technical, non-technical, action-based, and social learning. The system includes various training and development mechanisms. Additionally, they emphasize the importance of developing sustainability through ethical conduct, outcome metrics, effective organizational structures, action plans, employee training seminars, workshops, and other related methods. It is recommended that organizations prioritize sustainable practices in learning and incorporate diverse learning methods to enhance attention, participation, engagement, and knowledge sharing among learners. These efforts will contribute to more effective learning in Indian HE institutions.

16.6.3.2 TEACHING COLLABORATION

Educational institutions are encouraging teaching collaboration by engaging corporate professionals to share their knowledge through guest lectures, industry series lectures, and invitations to corporate events in the education sector. Corporations also have provisions to participate in sharing knowledge and expertise as part of Campus to Corporate programs, which are beneficial for students.

16.6.3.3 RESEARCH COLLABORATION

Research and HE go hand in hand in the development of a country. Only 4% of research expenditure in India is spent by universities; in contrast, 17% is spent by American universities, while it is 23% in the case of German universities. These numbers highlight the need to encourage research funding and research work in India. There is not much opportunity for Indian universities to collaborate with research centers, nor is there much networking between them. Furthermore, there is still a prevalent mindset that research is not an essential component in designing HE courses leading to increased job opportunities. China spends on 708 researchers per 1 million people, which is six times more than what India does. To address this, a Science and Engineering Research Board has been established to monitor research activities, resulting in a steep growth of contributions in research and publication in India. Indian authors have made significant contributions to research in recent years.

One report labels India as a "dormant giant that seems to be waking." At the same time, there is considerable growth in patenting activity. India Today (2005) reports that India is seen emerging as a global hub for commercial R&D activity, leading to an enhanced R&D scope for HE in India. However, there is still a shortfall of talent and resources in HE, hindering this development. Overcoming this shortage of quality researchers and research output will require expansion and wider opportunities for postgraduate research and doctoral studies in India. Holbrook & Dahl (2004) have shared that in order to reduce conflicts among faculty members with respect to their interests and to limit conflicts of interest and enhance their commitment, the institutes should consider revising their research policies.

Collaborative learning is an integral part of a collaborative culture. Encouraging collaborative learning is essential for fostering collaboration within and between classes, as well as across different courses. While collaboration in all its forms enhances learning in educational institutions, it is equally important to cultivate a collaborative culture among students and faculty. This type of collaboration contributes to improved learning outcomes for students. However, often the focus on the impact of collaboration is overlooked or neglected (Pertuze et al., 2010).

Communication plays a crucial role in collaboration, particularly in defining roles. Faculty and staff face challenges due to overloaded and under-funded initiatives, satisfaction surveys, strategic plans, increasing competition, and new data collection procedures. The role of faculty in establishing collaboration is not clearly defined, and there is a lack of research on the role of academic leaders and faculty in establishing networking and collaboration. Edmondson et al. (2012) emphasize the importance of academic leadership in building long-term strategic partnerships with flexibility (Calder, 2007). Academic leaders play a vital role in establishing rapport and support to create a collaborative culture and environment. This can be achieved by strategically sharing insights through appropriate media such as emails, notice boards, blogs, and more. Encouraging increased publications and citations also invites relevant collaborations.

16.7 CASE OF INSTITUTE OF PUBLIC ENTERPRISE (IPE), HYDERABAD

The Institute of Public Enterprise (IPE), located in Hyderabad, is renowned as a research institution. Its primary focus is on conducting continuous research on various issues related to the creation, implementation, revision, monitoring, control, and evaluation of policies concerning public enterprises. Over time, the institute has also extended its attention to private organizations. It receives partial funding from the Ministry of Human Resource Development, Government of India. Additionally, it has been recognized as Center of Excellence in Social Science research by the Ministry of Human Resource Development, Government of India, and the Indian Council of Social Science Research (ICSSR), another prestigious organization. The institute is well-known for its timely delivery of results and its strong emphasis on practical application.

16.7.1 GOVERNING BODY

The governance of the Institute is conducted through its Board of Governors, which formulates policies for the functioning of the Institute and periodically evaluates its performance. Having a proper governance structure and a prestigious board is a mechanism that enables collaboration, benefiting the Institute in terms of branding, placements, and networking in all aspects.

The elected/nominated members of the Board are granted a term of three years and are eligible for re-election/renomination. The elected members are chosen during the annual meeting of the General Body. The appointed members of the Executive Committee serve a term of three years and are also eligible for reappointment. The President is elected by the board for a period of three years (Table 16.3).

TABLE 16.3 Board Structure at IPE

Category	Details
President	Retired IAS official
Members	Key officials CMDs of leading CPSEs
Emeritus	Professor of eminence
Co-opted member	Academicians, key officials of leading private enterprises
Nominee	• Member Secretary, ICSSR
Members	• Secretary, DPE, GoI
	• Vice Chancellor, Osmania University
	• Principal Secretary to Government and CIP, Industries and Commerce, GoT
	• Principal Secretary to Govt, PED, GoT
Director of the institute	Existing director

Source: By authors.

16.7.2 RESEARCH AND CONSULTANCY AT IPE

The scope of research of IPE are majorly thrust into the following:

- Ministries of the GoI;
- SCOPE;
- Department of public enterprises (DPE);
- Planning commission;

- Comptroller and auditor general of India (CAG);
- Pay revision committees. IP.E.

IPE is also involved in journal publication. They have six in-house journals, including the IPE Journal. The IPE Journal is a refereed journal that focuses on public sector issues in India. Additionally, IPE is recognized for its involvement in consultancy assignments, serving clients from India and abroad.

Some of the reputed clients of IPE are (Figure 16.2):

- United Nations Industrial Development Organization (UNIDO);
- Asian Development Bank (ADB);
- Commonwealth Secretariat;
- United Nations Development Program (UNDP);
- Government of UK;
- International Center for Promotion of Enterprises (ICPE);
- World Bank;
- Department for International Development (DFID);
- Central and State Governments;
- Public and Private sector enterprises;
- Public and Private Sector Commercial Banks;
- Hindustan Copper Ltd.

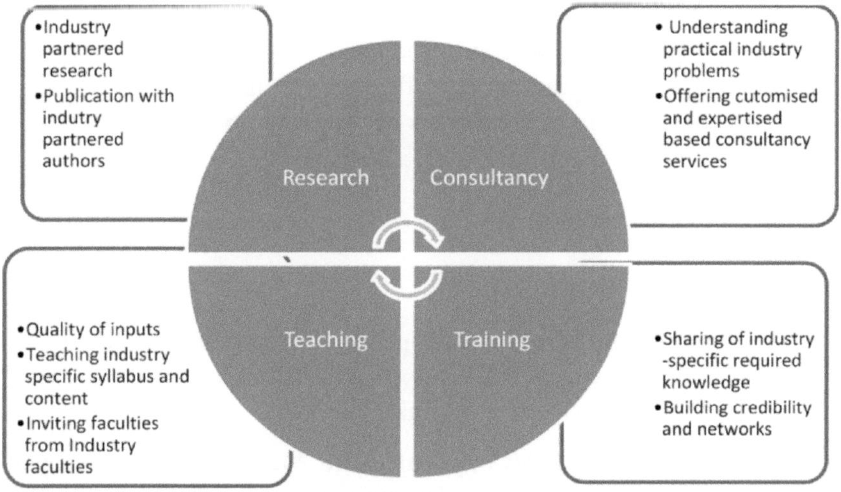

FIGURE 16.2 The collaboration model followed at IPE.

Source: By authors.

The research studies conducted by IPE have been extensively referred to by the Committee of Public Undertakings (CoPU), several ministries of the Government of India (GoI), the Planning Commission, SCOPE, the Department of Public Enterprises (DPE), the CAG, and pay revision committees. IPE publishes nine in-house journals, including the 'IPE Journal,' which remains a unique referred journal on public sector issues in India (Table 16.4).

TABLE 16.4 Dedicated Centers of Research in IPE

• Center for energy economics.	• Wage and salary studies cell.
• Center for state-owned enterprises and disinvestment.	• Center for innovation.
• Center for corporate governance.	• Center for governance and public policy.
• Center for corporate social responsibility.	• Center for sustainable development (CSD)
• Center for corporate risk management.	• Databank on central and state level public enterprises in India.
• Center for public enterprises and disinvestment.	• Case research center (CRC)

16.7.3 TRAINING

The Institute has been organizing Training Programs for the past 53 years for officers of the Public Sector, Private Sector, State, and Central Government undertakings. IPE has conducted nearly 100 programs for over 1,600 IAS, IFS, and other government officers. IPE has trained over 52,000 practicing managers in various types of programs over the years.

16.7.4 IPE MEMBERSHIPS

There are two types of memberships.

1. **Patron Membership:** It is open to corporations, companies, institutions, and organizations upon a one-time lump sum payment of Rs 1,000,000 (Rupees 10 Lakh only).
2. **Life Membership/Corporate Life Membership:** It is open to corporations, companies, institutions, or organizations upon a one-time lump-sum payment of Rs 6,00,000 (Rupees Six Lakh only).

16.7.5 ONLINE COLLABORATION DURING THE PANDEMIC TIMES

The pandemic has compelled academicians worldwide to embrace online learning (OLL). Within a short period, staff members have had to familiarize themselves with technology and remote learning, which they may have had little prior experience with. The closure of universities and academic institutions has significantly impacted the lives of both students and academicians, particularly in relation to their work, home, and social life. The usual academic processes were quickly replaced by inadequate alternatives, resulting in mediocre outcomes focused on meeting educational standards. Disruptions to academic activities such as lectures, assignment methods, and engagement in research and projects in HE has become inevitable during the pandemic. In fact, the challenging times have led to significant innovation in the education system, which would not have been adopted under normal circumstances. IPE has encouraged online internships with corporations and facilitated a 100% online placement process. IPE has provided necessary support to all parties involved and successfully fostered collaboration during the pandemic.

16.7.6 ALUMNI CONNECT

Taking advantage of the online platform during the pandemic, IPE organized several events where an alumni panel comprising alumni from all across the globe addressed the current students. In these events, alumni of IPE who are now working in various reputed organizations delivered talks on setting up their own companies or shared their experiences to enlighten the young minds. They narrated stories of their success and global work experience.

Another platform, the Hangout hole panel discussion, shared the best practices of student life as PGDM students and enriched the student community with the know-how of award-winning practices. They talked about their experiences at IPE and their journey post IPE. They used the platform to share with the audience the current business scenario in India and abroad.

16.8 CONCLUSION

Though there is significant growth and development in terms of the number of universities, colleges, Gross Enrollment Ratio, budget allocations by central and state governments, there is a lot of scope for improvement when compared to developed and developing countries and world statistics. Maintaining the standards of HE institutions and colleges should be motivated for quality assessment and accreditation by NAAC. Improving institution-industry partnerships for mutual benefit is key to a sustainable future. Leveraging collaboration to the advantage of an institution is an art and requires strong leadership. Building a collaborative culture is challenging as customary practices of working in silos are always comfortable for existing people. Enabling the right communication practices and framework is also vital in ensuring a difference in the workplace. Collaboration is successful only if all parties involved are equally involved. Many institutions have collaboration on paper rather than in practice, which is where senior leadership plays a strong role in ensuring impactful collaboration. The case of IPE, Hyderabad is an attempt to share practices in the form of a collaborative legacy.

KEYWORDS

- **collaboration**
- **collaborative legacy**
- **corporate collaboration**
- **higher education**
- **industry-academia linkage**
- **online collaboration**
- **qualitative research**

REFERENCES

Awasthy, R., Flint, S., Sankaranarayana, R., & Jones, R. L., (2017). Bridging the gap—A workshop of industry practitioners and university researchers. In: *TENCON 2017–2017 IEEE Region 10 Conference* (pp. 2504–2509). IEEE.

Borrelli, G., Cable, J., & Higgs, M., (1995). What makes teams work better. *Team Performance Management: An International Journal, 1*(3), 28–34.

Calder, E. S., (2007). *Best Practices for University-Industry Collaboration.* Ph.D. thesis, Massachusetts Institute of Technology.

Collaboration in Higher Education: New Directions for India and Australia. http://www.australiaindiaeducation.com/collaboration-in-higher-education.html (accessed on 20 February 2024).

Cyert, R. M., & Goodman, P. S., (1997). Creating effective university-industry alliances: An organizational learning perspective. *Organizational Dynamics, 25*(4), 45–57.

Edmondson, G., Valigra, L., Kenward, M., Hudson, R., & Belfield, H., (2012). *Making Industry-University Partnerships Work; Lessons from Successful Collaborations, Science Business Innovation Board.*

Ghosh, R., (2019). *Industry-Academia Linkage is Only 4.7 Out of 10 in India, Find out the Reasons.* http://timesofindia.indiatimes.com/articleshow/71684003.cms?utm_source=contentofinterest&utm_medium=text&utm_campaign=cppst (accessed on 20 February 2024).

Haugh, H., & Talwar, A., (2017). Sustainability – how do corporations embed sustainability across the organization? *Academy of Management Learning & Education, 9*(3).

Holbrook, K. A., & Dahl, E. C., (2004). *Conflicting Goals and Values. Buying in or Selling Out?* (Vol. 89). The Commercialization of the American Research University, Rutgers University Press.

Karamat, P., & Petrova, K., (2009). *Collaborative Trends in Higher Education, Systemics, Cybernetics and Informatics, 7*(2), 57–62.

Lai, E., (2011). *Collaboration: A Literature Review*. Pearson Research Reports.

Pertuze, J. A., Calder, E. D., Greitzer, E. M., & Lucas, W. A., (2010). Best practices for industry-university collaboration. *MIT Sloan Management Review, 51*(4).

Petrova, K., & Sinclair, R., (2008). Chapter VII – e-learning value and student experiences: A case study. In: Peñalvo, F. G., (ed.), *Advances in E-Learning: Experiences and Methodologies* (pp. 25–36). Hershey, PA: Information Science Reference, IGI Global.

Prigge, G. W., (2005). University-industry partnerships: What do they mean to universities? A review of the literature. *Industry and Higher Education, 19*(3), 221–229.

Rangarajan, S., (2021). *Critical Collaborations*. https://www.thehindu.com/education/partnerships-between-industry-and-academia-will-be-instrumental-to-advancing-research-and-knowledge-and-creating-a-skilled-workforce/article33702484.ece#:~:text=The%20Framework%20of%20Industry%2DUniversity,is%20preventing%20optimum%20co%2Doperation (accessed on 20 February 2024).

Reale, E., & Primeri, E., (2015). *The Transformation of University Institutional and Organizational Boundaries.* Sense Publishers: Rotterdam, The Netherlands; Boston, MA, USA.

Roschelle, J., & Teasley, S. D., (1995). The construction of shared knowledge in collaborative problem-solving. In: O'Malley, C. E., (ed.), *Computer-Supported Collaborative Learning* (pp. 69–97). Berlin: Springer-Verlag.

Roschelle, J., (1992). Learning by collaborating: Convergent conceptual change. *Journal of the Learning Sciences, 2*, 235–276.

Schneidewind, U., (2009). *Nachhaltige Wissenschaft. Plädoyerfür Einen Klimawandelim Deutschen Wissenschafts- und Hochschulsystem [Sustainable Science. Plea for Climate Change in the German Science and Higher Education System].* Metropolis: Marburg, Germany.

Index

C

F

G

H

Q

R

S

T

U

V